GARDENING WITH SOUL

Healing the Earth
and Ourselves with Feng Shui
and Environmental Awareness

GAYLAH BALTER

Learning Tree Books
Fayetteville, AR

Learning Tree Books
3038 N. Malinda Drive
Fayetteville, AR 72703
479-582-4145

Cover photo from Getty Images, www.gettyimages.com
Cover and interior design by Lightbourne, www.lightbourne.com

Publisher's Cataloging-in-Publication
(Provided by Quality Books, Inc.)

Balter, Gaylah.
  Gardening with soul : healing the earth and ourselves
with feng shui and environmental awareness / Gaylah Balter. -- 1st ed.
  p. cm.
  Includes bibliographical references and index.
  LCCN 2002 190312
  ISBN 0-9707861-1-5

  1. Feng-shui gardens. 2. Soul. 3. Sanctuary
gardens 4. Gardening--Environmental aspects.
5. Nature. I. Title.

SB454.3.F45B35 2003   712'.2
                                    QBI33-641

Printed in the USA on recycled paper

CONTENTS

PART 3   A NEW VIEW OF THE EARTH

GARDENING WITH SOUL

Healing the Earth and Ourselves with
Feng Shui and Environmental Awareness

*Dedicated to all gardeners
who take the time to experience
the connecting of their souls
to the soul of Nature.*

NTRODUCTION

Why is the meaning and expression of soul so important to us and to the life of the planet? This question passionately pushed me to write this book. Many of us are so overwhelmed by the pace and stress of life that we have lost touch with Nature and its capacity to restore, renew, and nurture the soul. Connecting with Nature is one way to restore what is lacking in the public and private fabric of our lives. The soul, spirit, or inner self clamors for the nurturing which can only come from a connection to the Earth that includes all living things, as well as wild and cultivated places.

Has soul departed from our lives, culture, and its many institutions? I believe it has. The proof of this is all around us, as we pour toxins and noxious chemicals on our land, in our air and water, condone criminal behavior, and go about our business as if nothing were wrong. In the *Seat of the Soul*, Gary Zukav says that, "by placing itself at the pinnacle of the hierarchy of life . . . and assigning maximal value to that position, humanity has created a world in which the exploitation of the environment appears natural." Since all life on our planet is interconnected, this viewpoint has affected our emotional, mental, physical, and spiritual health.

Our partnership with Nature is enriched when we find sparks of the Creator in everything around us. If this concept were widely accepted our world would become less polluted and the destruction of land, air, water, and species would cease. If you listen, you will hear your soul crying out for healing, wholeness, and balance in the world. My book aims to awaken in you a desire to join the work of setting things right in the environment and to become an Earth steward.

"This we know . . . the Earth does not belong to man; man belongs to the Earth. Man did not weave the web of life: he is merely a strand of it. Whatever he does to the web, he does to himself."

Attributed to
Chief Seattle,

My research has shown that Chief Seattle did not say the words attributed to him in 1854. It is believed that a screenwriter for a documentary about the environment wrote these words in 1972. I believe that the great Chief would have approved of them nonetheless.

Gardens feed our yearning to feel part of an ongoing process that has preceded us and will follow after our death. If we don't feel this bond with the natural order, we may find that Nature becomes an abstract, distant, and unfamiliar opponent that engenders fear and helplessness in us.

Feng Shui can help nourish this bond, because it asks us to look at land, gardens, landscapes, and our environmental decisions with a more spiritual and sacred view. Centuries of wisdom and soulful consideration of the environment form the core of this mindful practice. Every Feng Shui decision is concerned with creating harmony for people and the environments in which they live, work, and play. Honoring the energetic flow of water, air, and life force on a property is central to Feng Shui. Its holistic practices feed the soul and heal the body and mind because they restore balance and comfort to people's lives.

Every garden has the ability to rekindle our connection with Nature and provide ways to nourish the soul. I also believe the gardening community can help protect the environment from the incursions of toxic chemicals and unconscionable development. Carl Totemeier, a well-known horticulturist and Fayetteville, Arkansas garden columnist, feels that all gardeners need to become "frontline environmentalists."

Quotes, bits of wisdom, and gardening tips that have come my way as I have tended to my own soul are sprinkled throughout the book. Many years ago, after reading Rachel Carson's, *Silent Spring*, I concluded that everything we do in life is interconnected and that it all boils down to treating the whole world as if it were part of our own backyard. I would enjoy hearing from you and reading your comments.

—Gaylah Balter

$\mathcal{A}$CKNOWLEDGMENTS

I wish to thank the following people for helping me make this book a reality. Many people had a part to play behind the scenes giving me advice, gently forwarding comments and critiques, and letting me know they felt the book had merit. I could not have completed the manuscript without their help. I learned that it takes a village to write a book.

I am deeply grateful to Debbie Self whose editing skills and support helped to create a more polished book. Thanks go to Bob Haslam for his editing and encouragement to include more personal stories. Michelle Parks provided many editing and creative suggestions. I looked forward to my weekly visits to Dr. Roma Lisa Gray and to her sensitive and spiritual observations. My children, Anaya and Ariel, and Ariel's fiancée, Karen, supported me as they e-mailed their comments, advice, and critiques. A special thank you to my book designer, Shannon Bodie at Lightbourne, whose expertise and artistic eye make the book a joy to read. Gloria Pendry, my artist, produced remarkable drawings that made the ideas expressed come alive. Cynthia Morin deserves thanks for her thoughtful advice and help in re-organizing sections of the book. Linda Seamans at Paterson Printing deserves praise for working so closely with my desires to produce a special book. I cannot forget to thank my dog Joseph, who gives me encouragement in whatever I do. And to my bathtub, "Thanks for giving me a space to dream, create, and relax."

PART ONE

CHARACTERISTICS OF A SOULFUL GARDEN

"Man experiences himself, his thoughts, and his feelings as something separated from the rest, a kind of optical delusion of his consciousness. Our task must be to free ourselves from this prison by widening our circles of compassion to embrace all living creatures and the whole of nature in its beauty."

—Albert Einstein

GARDENING
WITH SOUL

"In a world moving at hyper-speed, where so many of us are anxious because of the rate of change, the soulful move is the move toward contemplating the source of things deeply rooted in eternity, the things that always are."

—Philip Cousineau

Gardening with soul encourages the discovery of your inner self and also some ways to garden the soil of your own soul. Such a view of Nature can lead you down paths of growth and enchantment while increasing your understanding of the Earth's resources and needs. I feel sure your garden will teach you more about yourself as you explore its unique place in your life and partake of the opportunities it offers to grow spiritually. As we care for tender plants and await the miracle of growth to take place we know deep inside how grateful we really are to have such a spiritual and mystical relationship in our lives.

We are given an opportunity to become aware of the sentient world around us each time we enter our garden spaces or for that matter anywhere in the outdoors and are reminded of the relationship that connects all beings to a divine source.

In a book about soul, it would seem advisable to explore the essence and significance of its meanings. Thomas Moore, Gary Zukav, and Anthony Lawlor have discussed the nature of soul and

A garden nurtures the soul because beauty and harmony are there and that is what the soul desires.

1

The soul is
not a thing, but
a quality or
dimension
of experience
that lives partly
in time and
partly in eternity.

spirit in their writings. I agree with them that our emotions, feelings, and spiritual experiences as well as the sacred, divine, and eternal aspects of life are all attributes and manifestations of soul. We extend our understanding of the soul's qualities to include the idea that it plants the seeds of truth and justice in the world and yearns for wholeness and connection to Nature. It is distinct from the physical body and many times describes a person's mettle, courage, and daring.

This discussion would not be complete unless we also understand the various meanings of spirit. Spirit represents the physical body and its vital functions, the life force in all living things, and it also infuses life with activity, energy, and ardor.

In some literature the terms *spiritual, innermost, sacred,* and *divine* are used interchangeably with spirit. As you can see these are also used to describe some of the attributes of soul. In this book I have made the decision to use spirit to express the life force in all living things and soul to express the spiritual, sacred, divine, and emotional aspects of life, Nature, and human pursuits.

The soul loves active participation with beauty and desires a space of quietness and reverie for replenishing the well of joy and happiness. It also requires a space that beams up sacredness easily and with relish. These are readily available in a garden.

WAYS TO GARDEN WITH SOUL

- It is wonderful to create a sitting area or bench in your garden where you can sit with other people to resolve conflict and discuss problems. This spot can become a niche for peacemaking; a familiar place to soothe wounds and be your own therapist. Prepare your special spot and nurture it.

- We enter the world of the soul when we give to ourselves and our gardens a thoughtful consideration of Nature as we talk to our plants, value even the most trivial of tasks, and reach for that conscious place inside of ourselves that has reverence for the Earth.

- Not imposing your will on your garden but letting it reveal itself to you is one of

Dr. Roma Lisa Gray's garden-sensitive philosophies that fosters a soulful approach to gardening. I planted some cleome seeds in areas where I desired them to grow. They did not do well there and instead came up in a spot that I would never have chosen. I stepped back and decided to enjoy them where they wanted to be.

Gardening with soul will enhance your sensitivity to life and you will become aware of the magical transmission that takes place between Nature and human. "With this sensitivity, trees take on a unique personality, flowers alert you to their essential being, animals and bugs show you their private lives, and birds include you in their play," says Judith Handelsman, author of *Gardening Myself, A Spritual Journey Through Gardening.*

Daily activities become paths to deepening your connection to the Earth as does caring for the environment. In your walk each day begin to notice something different or unusual about living things, such as the dry furry texture of the underside of a leaf and the smooth waxy quality of its other side. Then expand your experience by wondering what is the purpose.

There are many ways to become involved with the environment: wrap your arms around a tree; walk barefooted; feel the wind on your skin; listen to nature's sounds; go bird watching; garden; one day a month look at something in nature carefully, closely, and intimately; join an environmental group; go to your children's school and ask them to present an environmental program.

These simple yet profound acts valued by one person can multiply and then influence many other people and eventually change our world for the better. If each of us did one of these every month the tone and purpose of our society would begin to change and soul would again be part of social discourse and experience.

The soulful gardener cares for the needs of the Earth, thinking and acting as if God were in everything.

What is Gardening with Soul?

- A Belief that all living things have a spark of the divine and therefore are to be valued and protected.

- An enriching partnership with Nature as we learn to recognize and continue to find sparks of the Creator in everything around us.

- Creating sacred space where there is no sense of superiority over Nature but rather a feeling of interconnectedness of all things.

- Entering joyously into the flow of creation.

- Looking to Nature for its lessons and wisdom.

- A compassionate and nurturing view that sees water and soil and all life forms as having consciousness, intelligence, purpose and value.

- Talking to our plants with reverence and valuing even the most trivial of tasks in the garden.

- Holding rich, dark, fertile soil in our hands and contemplating all the promise that it has for life and for our future.

- Richly experiencing the Earth as she speaks to all our senses, satisfying us through her fragrances, sounds, textures, and other sensual and dazzling reminders of our passionate connection with her.

- New paradigms: weeds are beautiful, useful, and have a medicinal value and insects can be respected for their role in the balance of nature.

- Embracing the understanding that our environment reflects who we are, where we are going, and who we wish to become.

HEALING AND SANCTUARY IN THE GARDEN

"The earth . . . when we listen and watch and feel her flowing, then we remember deeply, the rhythm that sustains order and structure in life."

—Philip Sutton Chard

Healing is such a personal and transforming experience. We don't necessarily look for it as we dig in the garden; it just happens. Nature is subtle, full of surprises, and weaves enchantment and harmony into every leaf, flower, and particle of soil.

In our gardens we can establish a relationship with the Earth and also benefit from its abilities to mend and minister to our soul. We discover that stress, illness, and worry are alleviated by the peacefulness and vitality found there.

WAYS THAT A GARDEN HEALS

GARDENS RELIEVE STRESS AND ANGER

The garden fosters psychological health as it helps the soul restore balance and harmony.

We often forget that Nature is the prime source of the spiritual in life; block it out and we cut off our souls' source of nourishment.

5

- We can get rid of tension, mental fatigue, worry, and anxiety in a garden because it provides a quiet and peaceful refuge. Anger, rage, and resentment leave as we prune, dig, and do the hard work of gardening.

- Gardens decrease stress, renew energy levels, and generate positive feelings while providing a protected space where hope can flourish. After September 11, 2001, it was reported in the media that more people were sitting in public gardens and parks. Amid the horror, gardens provided relief. Carl Totemeier, a horticulturist, has reported on this phenomenon as well. He tells the story of a couple who years ago came to the Botanical Gardens on Long Island where he worked. They sat together on a bench in the gardens after their business failed and found the courage and inspiration to build a new one. And they succeeded.

- In the garden we learn to accept what happens, make adjustments, and go on. Many believe that for healing to take place we have to surrender and let go. What better place to do this than in a garden? In fact, this occurs naturally as we garden.

GARDENS INCREASE WELL-BEING

There is a message for us in any illness; we learn from observation and acceptance, not by resisting.

- We discover that the many subtle nuances of life and death experiences that occur in a garden also exist in life; this helps us find renewed strength to face sadness, trauma, crisis, and loss in our lives.

- According to the bushman of the Kalahari, we are all born with the seeds of disease within us. When our lives fall out of balance, these seeds find fertile ground to grow. We look around at Nature and stare straight into the face of balance. In our gardens we know that when the soil, air, sun, water, nutrients, and temperature are in balance, our gardens prosper. The garden gives us a way to re-create the balance we so desperately need in our lives, and shows us that balance is needed for healing.

• Gardening accomplishes much of what your doctor prescribes without using medicines. Many people notice their energy levels rise, blood pressure rates decrease, breathing deepens, and healing chemicals begin to flow. This improves the immune system and also causes the production of seratonin and endorphins to increase which then raise levels of pleasure, joy, and happiness.

• Doctors know that laughing is good for our well-being and a garden inspires humor and whimsy. (See chapter on Humor and Whimsy.)

• Nursing homes and elder day care centers have started gardens that produce lower death rates, increase the feeling of community, encourage higher levels of participation in activities, and improve the quality of life for members.

GARDENS RESTORE INNER PEACE

Gardens teach us to appreciate the present moment and trust that what happens is for the best.

• Restoration of communication with our hearts' language occurs in a garden. The tranquil sounds of nature encourage us to tune into our inner voice. We become more sensitive to our feelings, our hearts open, and we reach that place where honesty with ourselves and others paves the way to healing.

• Hope flourishes in a garden as you look down at the rich, dark soil and visualize plants that will emerge and the pleasure they provide.

• The garden beckons us to peel off the armor of protection that we have created around ourselves. Running away from our

7

In a garden you become an artist using Nature as your canvas. Your life is woven into the garden forming a tapestry that reveals your dreams.

emotions increases stress. It is better to face them and then soothe and heal them in a garden.

- Gardens teach us about finding the beauty that lies deep inside all of us. The real self is revealed and healing can begin. A garden is undeniably the place to find the Self.

- Moving beyond the habitual patterns of behaving, working, and thinking can change the course of your illness and life. In the garden we are constantly asking how we can adapt, think creatively, and change.

- All the lessons of life are available to us in a garden as we find restoration of our inner life, sustenance for our soul, and freedom to meditate and pray. Many feel the call of the garden to be as strong as the beckoning of religion or churchgoing and find that garden work is a form of prayer.

GARDENS INCREASE OUR ABILITY TO EXPRESS LOVE

Loving your garden and sharing that love with others builds up good feelings in a place and is the perfect antidote to a world filled with conflict and stress.

- Beautiful memories soothe our troubles and make us feel lighter and more hopeful. Cuttings or transplants taken from the garden of a relative or friend can create a sense of connectedness, and help alleviate feelings of separation and loss. A garden connects one generation to another when we lovingly remember those who have gone by filling our gardens with the plants they nurtured.

- People are drawn to a garden by their emotions, memories, and senses. Put in your garden plants that remind you of pleasant sites you have visited on your travels. The more ways you incorporate positive energies into your garden the more effective your healing experience will be.

Nature provides ways for us to heal many of our ills, if we search. Philip Sutton Chard, a psychotherapist, uses Nature as a means of healing for his patients. He encourages them to work out their issues by observing, walking in, and consciously cultivating a relationship with Nature. The wind listened to their anger, frustration, and bitterness and then whisked it all away.

GARDEN SANCTUARIES

Our culture lives in denial of the non-rational, non-verbal, mystical, and spiritual ways of knowing and elevates science to a religion. Other ways of knowing enable us to become less attached to the allure of science. Many people have reported complete healing, reversal, and remission of disease after praying, meditating, or using visualization techniques. Your garden sanctuary provides a place to do this.

A garden becomes our sanctuary because it offers us a place for rejuvenation and renewal and is a welcome vacation from the workaday world where our time is our own and we can express ourselves freely.

When your breathing slows and your shoulders relax, you have found your spot. Create a labyrinth, a bench, a niche, or a room that can become your sanctuary or retreat, and go there to tend your soul. The results will amaze you.

The Sanctuary Garden by Christopher Forrest McDowell and Tricia Clark-McDowell describes ways to create sacredness and sanctuary in your garden. You can register your garden with them as part of the Cortesia Sanctuary Project, which offers a registration packet that includes signs designating your property as a sanctuary. You may write to them at the Cortesia Sanctuary Project, 84540 McBeth Rd., Eugene, OR 97405.

There is beauty in aging, maturing, and ripening as Nature so wonderfully exhibits.

9

CREATING A GARDEN RETREAT

We look forward to spending time in our gardens after a long day at work. It is there we can ease the day's stresses. On weekends we dream, gaze, read, and entirely lose ourselves in the garden. At other times we join with friends and family, eating, and playing. As we regenerate our body, mind, and soul, we discover that joy emanates from the satisfaction that tilling the Earth brings. There is something decidedly magical and restorative about putting a seed, bulb, or seedling in the ground and then watching as it grows to either feed our body or a hunger for beauty.

WHAT MAKES THE GARDEN A SACRED SPACE?

- The garden gives us direct contact with the order and spiritual nature of the universe.
- It nurtures all aspects of the soul and the spirit.
- It allows us to steward the earth.
- Our gardens become sacred through purity of intention.
- The garden provides ways to access other realms of consciousness.
- The garden offers us glimpses into the deeper aspects of spirituality.

THE LABYRINTH: The facts and ideas that follow came from: *Better Homes and Gardens: Simply Perfect Garden Rooms*, 2001, "A Walk To Inner Peace" by Joanna Wolfe; some are from Sig Lonegren's book, *Labyrinths, Ancient Myths and Modern Uses*; and my own personal experiences and research.

A labyrinth is a type of garden room that you walk in rather than sit in. Its purpose is to awaken your inner wisdom while providing ways to heal. A specific spiral-like pathway is delineated on the ground or space you have designated. The walker follows the pattern laid out by going to the center and then turns to retrace his/her steps back to the starting point. (See the diagram on facing page.)

Labyrinths are an ancient tool for personal, psychological, and spiritual growth, healing, and transformation. Pictographs of labyrinths have been found in North America among Hopi ruins, India, Crete, Spain, and many other countries around

Many cultures use medicine wheels and mandalas in the garden to foster healing and spiritual journeys.

the world as well as in ancient Celtic rock carvings and amulets. Apparently every village in England had a labyrinth on the village green. In the 13th century, a labyrinth was installed in the stone floor of the Cathedral of Chartres, France. Many Cathedrals built in the Middle Ages in Europe also have them.

People have lost touch with something greater than themselves. The labyrinth serves as a connection to eternal truths and also provides healing for many issues. People have been healed of eye problems, Parkinson's disease and cancer. Some find after walking the pathway that they are more able to focus on their problems and find solutions; others ask for help and get answers; and still others emerge feeling more peaceful and relaxed. It works differently for each person. Some sources say the labyrinth represents our journey through life, life itself, or our search for spiritual meaning. All agree that life is like the labyrinth; it is unpredictable and takes many turns.

The right and left turns restore a sense of equilibrium that seems to be lacking in our lives. The winding path forces us to stay focused on the physical level and encourages us to relinquish control. Letting go and trusting creates a space to heal.

Labyrinths are popping up all around the country, in churches, homes, and healing centers. People are looking for more tools to reduce stress, heal disease, and find balance and harmony in their lives. They are turning to this millennia-old tool for inner peace and healing.

In the early 1990s, a replica of the Chartres labyrinth was installed in San Francisco's Grace Cathedral, initiating a new wave of interest in the labyrinth. More than 650 portable or permanent labyrinths are found in 49 states. You can purchase a labyrinth painted on a canvas to put wherever you want on your property or build one from plans.

I can hardly wait each morning to see how the garden has changed overnight nor can I wait to greet my over-wintering plot of land as spring arrives and brings with it the opportunity to begin gardening all over again.

Many are constructed as brick paths; some are just a colored cord laid out in the labyrinth pattern. Many choices are available.

In sacred spaces there is no sense of superiority over Nature; there is a feeling that the interconnectedness of all things is respected. We are nestled in a protected private oasis that connects our past, present, and future. When we are in the midst of a soul-nourishing activity such as sitting in a garden sanctuary or retreat, walking a labyrinth, or gardening we are unaware of time, connected to the source of love and wholeness in the world, and are more able to heal ourselves.

SYMPTOMS OF INNER PEACE by Saskia Davis

- A tendency to think and act spontaneously rather than on fears based on past experiences.
- An unmistakable ability to enjoy each moment.
- A loss of interest in judging other people.
- A loss of interpreting the actions of others.
- A loss of interest in conflict.
- A loss of the ability to worry. (This is a very serious symptom.)
- Frequent, overwhelming episodes of appreciation.
- Contented feeling of connectedness with others and nature.
- Frequent attacks of smiling.
- An increasing tendency to let things happen rather than make them happen.
- An increased susceptibility to the love extended by others as well as the uncontrollable urge to extend it.

WARNING: If you have some or all of the above symptoms, please be advised that your condition of inner peace may be so far advanced as not to be curable. If you are exposed to anyone exhibiting any of these symptoms, remain exposed only at your own risk.

Soulful Qualities of Trees and Plants

"The most beautiful thing we can experience is the mysterious, it is the source of all true art and science."

—Albert Einstein

This chapter is dedicated to the awesome and spiritual nature of trees and plants.

Trees appear repeatedly in all of the world's religions, literature, and folk tales. They symbolize all that is vital, wise, and eternal in our culture. Adam and Eve and the Tree of Knowledge are powerful icons for western civilization. The image of the Tree of Life emerges and with it the symbol of roots reaching deep into Mother Earth, and being secure and steady. The branches reach to the sky and symbolize fatherly care, concern, and protection. These combine in the tree and represent the integration of all cosmic energies.

Trees have been revered by all cultures through the ages. Nature was not a separate entity to these societies; it was intricately linked to every part of their lives. Sacred groves of trees have been preserved in every corner of the globe. Historically people recognized the spiritual value of trees and understood that their presence aided clarity, fertility, and a connection to the divine. Trees symbolized knowledge and wisdom, and were considered teachers, vehicles of prophesy, and protectors of the sacred. They are a timeless and archetypal image of rebirth, growth, vitality and security.

We have turned our backs on this ancient wisdom and now our culture uses trees as throwaway items. We think of them as so many board feet. Ronald Reagan, as governor of California, defended the mentality of the timber industry and its rights. He said, "A tree's a tree. How many more do you need to look at?"

Many helpful drugs come from the bark of trees: aspirin, from willow bark; yew gives taxol for cancer treatments; pine is known to be a powerful antioxidant; the elm offers lozenges for sore throats; and the birch has anti-tumor properties, to name a few. As we chop down trees, we lose many possibilities for healing ourselves and the planet.

The following list describes some of the many ways that trees give us sustenance.

- Trees provide nesting sites and protection for birds and animals.
- Trees give off life-sustaining oxygen and clean our air.
- Trees renew available water to the atmosphere and create rain-producing clouds while modulating the temperature and climate.
- Trees provide watershed protection and anchor the soil preventing mudslides and flooding.
- Trees provide inspiration and beauty for us all.

Some things we don't yet fully comprehend about trees; one is that we don't give them credit for having intelligence. An oak tree under attack by insects sends out messages that alter its chemistry to produce compounds to deter the onslaught. Other trees nearby not yet experiencing problems begin the same chemical defense, as if responding to a call for help. Some scientists do not believe this has been totally proven. I found this information in several books that appear in my bibliography and present it here to give you other views of Nature to ponder, as I do throughout this book.

Many books describe how trees increase the vitality and energy of an area. Some believe they serve as antennas and that energy spirals in their trunks. Native Americans replenish their energy by putting their backs against a tree. In workshops

Our future depends on trees. How can we allow the indiscriminate cutting down of the rain forests that provide our oxygen and rain?

"The seat of the
soul is the heart."

Thomas Moore

I have taken, we were asked to do this and found we were refreshed.

Trees are ancient survivors of life on this planet and serve as record keepers. The redwoods of California existed when Jesus lived. I lose count of the ways that we mistreat them. Sensitive and wise individuals have devoted their lives to the preservation and care of trees. Legendary Johnny Appleseed and Arkansas native Julia Butterfly, who recently lived for two years in a California redwood she called Luna, come to mind. Others like Elzeard Bouffier have changed the Earth through their devotion to trees.

Where the Alps come down to meet Provence, France, lush valleys, villages, forests, and streams once existed. It then became a barren and lifeless stretch of land. The people had moved away abandoning their villages.

One man changed all that. After losing his only son and wife during World War I. Elzeard Bouffier devoted his life to the reforestation of the region and decided that planting trees would become his daily work. Forty years of planting more than 100,000 oaks and other varieties by hand each and every day and living in a hut in the area, changed the climate, weather patterns, water flow, and ability of the soil to support life. One can go there now and see that people have moved back to the once empty villages. Life flourishes there anew, streams have started to flow again and the soil has become more favorable for crops and life. The countryside glows with vitality. Trees planted by " . . . an old and unlearned peasant who was able to complete a work worthy of God," changed all that. *Country Living*, March 1997. From *The Man Who Planted Trees* by Jean Giono.

There is a modern-day Johnny Appleseed. Tim Womick travels all over the United States, teaching, singing, and storytelling about the importance of trees, the environment, and the role of trees as Earth stewards. It is a project sponsored by the National Tree Trust, Washington, D.C., 800-863-7175.

"A spiritual essence lies behind the material form of a plant."

Goethe

Many herbal remedies are ancient and time-honored sources of healing, relaxation, and nourishment.

I cringe and cry out when a tree is cut down. It hurts me deeply. Somehow they feel human to me and I sense their pain. When I moved to my property in Fayetteville, Arkansas, more than five years ago there was a forlorn ash tree in my backyard. All who came to work or visit said I must cut it down. Ants crawled up and down its trunk, several gashes pierced its bark, and a sizeable wind might push it over. I decided to use another method of care. I mixed up a solution of Bach Flowers Rescue Remedy, poured it over the ground around the tree, and said some prayers and words of encouragement. In about a month I noticed a profound change in the ash. The gashes were healing, the ants had vanished, it had sprouted new leaves, and it somehow appeared stronger and healthier. It is now one of the most attractive elements of my garden. I often go over to this lovely being and whisper words of praise and continued wonder about the miracle that took place.

PLANTS

Plants have a purpose as part of the interconnected, dynamic relationship between humans and all other life forms. Many species become extinct everyday so I am wonderstruck that new plants are being discovered daily in South America, Africa, and Asia that hold promise for healing disease.

In *The Secret Life of Plants*, authors Peter Tompkins and Christopher Bird provide ample evidence that plants have nervous systems analogous to ours. Plants were observed and their reactions to events were measured by a galvanometer. Researchers report how Bach and Mozart played near plants created greater growth. In an experiment at the University of Indiana, seeds sprouted earlier, and stems grew thicker and greener when Gershwin's "Rhapsody in Blue" was played near plants. Plant reactions to the death of living tissue were also recorded, as were responses to other events and experiences. Tompkins and Bird tell of other valued studies which prove that sentience exists in all living things down to the subatomic level and that plants have nervous systems. It is important to note here that some people do not agree with their findings.

SOME NEWS ABOUT PLANTS

Wes Jackson reports in *Secrets of the Soil,* also by Peter Tompkins and Christopher Bird, that scientists have perfected a type of wheat that is high in protein and produces such an enormous amount of seed that it would save farmers much labor, time, and money. This illustrates the elegance of plant simplicity and the absolute proof of the overflowing abundance that exists in Nature, while providing a real contrast to the seeds Monsanto sells which produce sterile crops.

For information on this topic and others go to www.organic-consumers.org.

The plant kingdom can help extract radioactivity and other pollutants from soil and sewage. Sunflowers, mustard seed, and bulrushes are some of the many plants being studied that do this. The city of Arcadia, California sends their waste through a bog containing some of these plants. They have discovered the water from this system emerges pure enough to drink.

Plants are also being used to remove air-borne toxins in homes and offices.

- Boston Fern, spider plant, and philodendron eliminate formaldehyde that comes from foam insulation, plywood, carpeting, clothing, paper goods, furniture, and household cleansers.

- Bamboo palm, corn plant, mother-in-law tongue, and many other plant varieties help to purify our indoor air and bring peace and calm to busy spaces.

When picking or pruning plants it is good to ask who wants to come and gently tug to see if the plant comes easily. If it doesn't, go to another and ask again, reports Judith Handelsman. Allow yourself a few

"Every blade of grass has an angel over it, saying, 'Grow'."

Talmudic saying

17

minutes of quiet to tune into and link up with your plants. Ask them where they want to be in your garden, near which plant. Don't worry; no one will think you are crazy if they hear you talking to your plants. They will just assume you are a bit eccentric.

Handelsman also suggests that we prepare a plant for transplanting 24 hours before re-potting it and also to tell your trees, bushes, and plants that you plan to do some trimming the next day. This allows the plant to prepare itself for the shock of being uprooted or cut. I say "thank you" when I pick plants for food or decoration because they are giving themselves in a completely selfless way and are like old friends, coming up year after year.

Put yourself in the position of plants as living things and ask how would you like to be treated. Imagine someone suddenly cutting all your hair off without asking permission first.

WILDFLOWERS

In 1900, women in Boston who wanted to become advocates for wildflowers started the Northeast Wildflower Society. They had learned that some wildflowers were in danger of extinction. There was such a craze for certain wildflowers that entire fields were being picked clean to sell in the Boston markets. These women recognized that the gems of the fields and forests were endangered. The society was formed to preserve them for future generations.

In Texas, wildflowers are a state treasure. Many states plant wildflowers on the median strips of highways, and as we drive by we are given a showstopper by nature. Many people have remarked how it makes driving more relaxing and easier.

GROW YOUR OWN MEDICINES AND REMEDIES

- **Ginger:** Relieves nausea, and acts as a digestive aid and gentle nervous system stimulant.

- **Cayenne:** Stimulates general circulation and relieves pain when applied externally.

- **Cinnamon:** Aids in digestion, kills parasites, and improves circulation.
- **Bilberry:** Is good for eyesight and constipation.
- **Blackberry:** Contains large amounts of vitamin C.
- **Cranberry:** Is used to heal kidney infections.
- **Raspberry leaves:** Soothes and tones the stomach, helps relieve menstrual problems, and tones the uterus.
- **Strawberry:** Provides vitamin C.
- **Currant:** Acts as a diuretic and helps to lower blood pressure.
- **Garlic:** Aids digestion and lowers blood pressure.
- **Chamomile:** Calms and aids sleep.
- **Sage:** Counteracts greasy foods.

In former times every homeowner had a medicinal herb garden. Here are Martha Washington's selections:

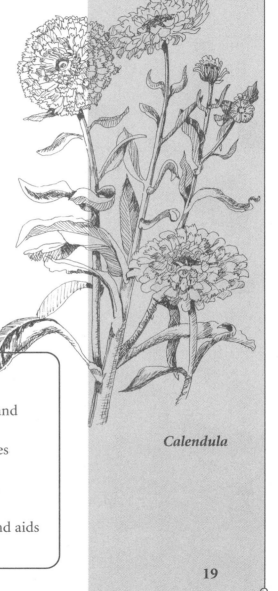

Calendula

MARTHA WASHINGTON'S MEDICINE CABINET

Calendula: Relieves muscle spasms, aids the healing of sores, and soothes the skin.

English ivy: Eases colds, diarrhea, and gastritis.

Garlic: Curbs parasites and reduces the effects of snake and spider bites.

Horseradish: Curbs parasites and eases sciatica.

Lavender: Eases headaches, relaxes, and encourages sleep.

Rhubarb: Curbs parasites and relieves constipation.

Rose: Eases headaches and improves digestion.

Thyme: Prevents wound infection and aids respiratory ailments.

TWO RENOWNED HEALING GARDEN ESSENCES

Wildflowers bring us messages of hope and the pure celebration of life.

Elderberry

Everything about this plant is useful, from root to twig, and has been touted as a cure for the common cold and the flu. It is found in herb stores as a liquid or in capsule form and has been used as an ingredient in liniments and salves and as an antiseptic, diuretic, and purgative. Research indicates elderberry also has antiviral and antioxidant properties. Some gardeners swear it wards off caterpillars. Make a strong elderberry leaf tea and spray it on plants threatened by leaf-munchers.

Echinacea

World-renowned author and herbalist, Stephen Foster, says, "Echinacea is nature's immune enhancer." He believes that it does not cure but seems to activate body tissues to initiate a healing process and stimulate the immune system. It also lessens symptoms of a cold or the flu and reduces the amount of time symptoms linger. Five of the nine species of echinacea are native to Arkansas, where Foster makes his home.

AROMATHERAPY ❧ BACH FLOWER REMEDIES ❧ HOMEOPATHICS

Aromatherapy:

Many flowers, herbs, and other types of plants are used as therapy. This knowledge is derived from an ancient craft that captures the soul essence of the plant and distills its oil for use in many healing techniques. It is said that everything that grows on this planet plays a part in helping humans to heal disease and pain. I have used aromatherapy for years to ease aches, pains, water retention, decreased energy, and emotional states.

Bach Flower Remedies:

Dr. Edward Bach, a London homeopathic doctor, was spiritually guided to find wildflowers that could remove the cellular memory of certain mental and emotional states and stressors, thereby allowing the body to heal itself. He created 38 remedies.

Each contains the energy signature of a specific plant. Since then many other individuals and groups have discovered the healing properties in American wildflowers, herbs, and even vegetables. The Bach Flower Rescue Remedy is specifically used to soothe trauma, shock, and injury in animals, humans, and plants.

Homeopathic Medicine:

Homeopathic medicine has developed over the last 200 years to become a vital source of healing in this country and around the world. It also uses the healing properties contained in plants and plant substances by distilling and potentizing their vibrational imprints. Using the theory that "like cures like," similar to the use of vaccines, they work with many diseases, everyday complaints, emotional states of mind, as well as poison ivy, itching, coughing, and nausea.

We all know in a place deep inside of us that plants have healing properties, yet our culture goes about the business of destroying their habitat. At this very moment, millions of acres of forests, prairies, marshlands, bogs, and valleys are being destroyed all over the world. Some of the plants that may hold the cure for lupus, cancer, heart disease, AIDS, and other diseases disappear forever. Our silence as this happens is one of the world's greatest tragedies.

I find something I like in every plant I see.

> *"I need flowers the way I need food. They make beauty tangible. Being in the company of plants is a tonic for the soul."*
>
> Leslie Goldman
> *Enchanted Garden*

 Garden Gift: **EDIBLE FLOWERS**

These flowers can be used in salads, for edible decorations on cakes, and in omelets or muffins: calendula, carnation, mum, cornflower, daylily, hibiscus, honeysuckle, Johnny jump-up, violet, rose, pansy, nasturtium, and scented geranium.

SOULFUL QUALITIES OF WATER AND SOIL

"Matter contains consciousness."   —Isaac Bentov

All life forms have consciousness, intelligence, purpose, and value. Some scientists, writers, and explorers of the unusual in Nature have uncovered ample proof of this by observing and believing that Nature has consciousness and is able to communicate with us. Many researchers have become aware of the various qualities of water and have written and lectured on their discoveries to limited audiences. These ideas may seem strange to you, but I feel they hold a key to understanding and appreciating our world. Many useful and commonplace discoveries initially seemed outrageous and nonsensical to us.

WATER

As water flows from mountaintops to the sea, it carries evidence of our collective irresponsibility. Every year chemical contamination and water-borne diseases kill 15 million people. Some of this pollution comes from human waste, some from agricultural waste, and some from industry. One in six humans must carry this precious liquid long distances from its source to their homes. *Water: The Drop of Life*, PBS Series. October 2000.

Is a river just a lot of water, or is it a reservoir of living energy that influences our lives on every level?

THE QUALITIES OF WATER: (Some of these are from *The Sanctuary Garden*, some are from my own observations, and others are from my resources.)

Water symbolizes everything that is fresh, clean, and renewed. It transforms all that it touches, is where life first emerged, and has long been associated with abundance and the ability to sustain life. Most ancient cultures believed water symbolized the feminine, the womb, fertility, and prosperity. It can provide a pleasant background sound obscuring harsh noises caused by traffic. The calming and tranquil properties of water provide a healing and peaceful environment in which to reflect and meditate. When water is present wildlife is attracted to the garden and it gives them sustenance.

The natural beauty of waterfalls and ponds offer a showcase for garden art, bring drama, and create a focal point in a space. Water epitomizes fluidity, movement, and grace.

Water also has many mystical, metaphysical, and spiritual qualities as shown by the following information. You will find that many in the scientific community do not agree with these research findings. I offer them here to increase your awareness of the many theories about water and its characteristics.

WATER RESEARCH

Masaru Emato, a water researcher from Japan, has experimented with water and discovered that it responds to prayer, good intentions and meditation, and also carries the effects of negative events and experiences. As people prayed nearby, the water in a polluted lake cleared and the crystals of water, shown by his photographs, changed from formless masses to beautifully configured star shapes. Emato demonstrated how our thoughts affect the structure of water. The water revealed how it reflects the consciousness and thoughts of people, has intelligence, and the highest retention capacity for information of any matter on earth.

Victor Shauberger, a 20th century German scientist and inventor observed spiraling water in a stream and discovered that it cools itself and produces energy by this action. His research shows water to be an accumulator and transformer of the energies originating from the Earth and the Cosmos, and as such was the foundation of

Emato,
"Water is a mirror of our mind."

all life processes. We fail to view water as a living entity, as he did, and have overlooked the potential for its creative uses.

Shauberger, Alexander Schwenk, and others provide a large body of evidence that some water originated deep within the earth, yet outside the core. Drilling into hard impermeable rock reveals an abundant supply of clean water. Primary or virgin water, as they call it, is a theory that certain rocks deep in the earth held hydrogen and oxygen gases that combined to form water. Shauberger and Schwenk also postulate that more water lies under the earth's surface than above it.

Tompkins and Bird write in *Secrets of the Soil* about farmers and gardeners who achieved improved growth in crops by first stirring water one way and then another and then flicking or spraying it on crops or individual plants.

In Eureka Springs, Arkansas, Barbara Harmony researches and lovingly disseminates valuable information about the varied properties and uses of water. She has established The National Water Center, gives workshops, and writes articles and books on the nature of water and its value. (See Resources-The National Water Center.)

Thinking less conventionally and using some of these researchers' knowledge would help the world rid itself of pollution, poor harvests, and increasingly shorter water supplies. Water is important to everyone, yet we treat it so contemptibly.

SOIL

Where would we be without soil? All life depends on its ability to provide nutrients and minerals to plants. Yet we allow it to blow away, erode into streams and rivers, and become so filled with assorted chemicals that it has become deficient in everything humans require for healthy bodies. The United States Department of Agriculture has repeatedly published data proving that the soil in our country is sadly

Gardens connect you to the soulful longings of your heart.

"Enchantment equals a state of rapture and ecstasy in which the soul comes to the foreground and the concerns of survival and daily preoccupations at least momentarily fade into the background."

Thomas Moore

depleted of nutrients and that topsoil levels are dangerously low. The shortsighted farming methods employed by the agribusiness over the past 60 years has helped cause this alarming situation.

There are many things the average food consumer can do to change the downward slide of our soil quality. Buying more organic products, growing your own food, and creating healthy soil right in your own backyard are good first steps.

COMPOSTING

Compost supports the health of the soil. It does not contain every one of the nutrients that healthy plants need but does contain ample supplies of calcium, copper, zinc, boron, iodine, manganese, and nickel. These are nutrients we look for in our vitamin pills because they are missing from our soil and from the vegetables grown in that depleted soil.

Ideally the compost pile should contain equal amounts of green garden waste and soil in well-moistened layers. Adding starters that help to heat up the pile will get things going faster. It is not a good idea to put in kitty litter, animal droppings, diseased plants, weeds, meat, grease, fat, or bones. Chemically treated wood products or sawdust from varnished or polyurethaned floors will release toxins into the soil that become part of the vegetables you harvest and eat.

THINGS YOU CAN SAFELY PUT IN YOUR COMPOST PILE:

- kitchen scraps
- garden refuse
- grass clippings
- straw
- shredded leaves

- dairy products
- egg shells
- moldy flours
- shredded paper
- sweeper bags

- coffee grounds
- tea bags (take the staple out first)
- bread

If we didn't have decay there would be no more room in the world. Decay creates life and soil and then feeds plants and gives more life back to us. It is also called black gold due to its value. Humus is a term used to describe decomposed organic portions of soil. Microorganisms break down this mixture into a friable, sweet-smelling, black soil. When it is crumbly, brown or black, smelling sweet, and earthy looking, it is ready to use anywhere in your garden. Compost can be used as an additive to help sandy soils retain moisture and clay soils improve drainage. Apply sparingly on herbs or you will get a lot of growth without the essential oils.

THREE SIGNS OF A PERFECT PILE:

1. **STEAM:** Heat is a sign that the microbial community is hard at work.
2. **A SWEET EARTHY SMELL:** Your pile needs oxygen. If it has too much nitrogen and not enough moisture it will not have a sweet earthy smell. It is important to keep the pile moist (like a wrung out sponge) and turn it over frequently.
3. **VOLUME:** To heat your pile properly you will need at least 1 cubic yard of waste which is 3 feet wide by 3 feet long by 3 feet high.

BENEFITS OF COMPOST:

- Improves drainage.
- Helps retain moisture.
- Improves the texture of the soil.
- It teaches children the benefits of organic gardening.
- Helps our environment because this waste would normally end up in our already overburdened landfills.
- It's free.
- Frees up minerals and nutrients that are already in the soil.
- Helps prevent and suppress disease.
- Allows air and water to move more freely in the soil layers.
- Brings earthworms and other beneficial life forms to the soil.
- Beneficial organisms already in the soil are activated and put to use in the garden.

Many cities have passed legislation that prohibits putting garden waste in the landfill, and they provide some form of pickup for this. When added to our landfills

27

Studies have shown that plants grown in soil mixed with compost resist insects and diseases and withstand frost better.

it promotes the buildup of methane gasses dangerous to the surrounding population, damages the ozone layer, and greatly contributes to global warming.

Underneath our very feet lies one of Nature's wonders, a whole world of organisms, insects, creatures, and other forms of life. It's hard to imagine all that activity going on without actually seeing it. Holding rich, dark, fertile soil in your hands and contemplating all the promise it has for life and for our future is a soulful experience. Creating and maintaining a compost heap in your yard is one way to become an Earth steward. As we become aware of the abundance and rich possibilities that organic gardening with our own compost brings, we connect with the Earth and appreciate her nurturing ways.

© Gloria Pond

🍃 🍃 🍃 *Garden Gift:* 🍃 🍃 🍃
HOUSEKEEPING WITH HERBS

Moth chasers are pennyroyal, cedar shavings, lavender, or rosemary. Add a few drops of essential oils to your laundry when washing sheets and towels and experience the fragrances of the garden as you sleep and prepare for your day.

MORE TIPS FOR A SUCCESSFUL COMPOST PILE
from Steven Cline, *Fine Gardening*, Sept./Oct. 1994

1. Lightly wet down each layer as needed to make it about as moist as a damp sponge. Green materials may not need any additional moisture.

2. Alternate layers of brown and green materials, making sure each layer is no more than a few inches thick.

3. Bury kitchen scraps in the bottom or center of the pile where animals won't dig for them.

4. Sprinkle a little topsoil or other compost between the green and brown layers. The topsoil settles the pile down, adds more microbes to the mix, and helps retain moisture.

5. Remember brown layers may contain leaves and twigs while the green layers may contain garden waste and grass clippings.

6. It is easier to maintain a pile in the shade rather than in a sunny spot.

COMPOST RESOURCES:

Your state's Cooperative Extension Service and ATTRA (see Resources) have information on composting, testing your soil and Master Gardener classes.

Books to consider for further reading:

Let It Rot by Stu Campbell

The Complete Book of Composting by J.I. Rodale and Staff

Backyard Composting, Missouri Botanical Garden, Kemper Center for Home Gardening, P.O. Box 299, St. Louis, MO. 63166: 214-577-9440. $12.50, for a 26 minute video.

COLOR

"If nature is your teacher, your soul will awaken." —Goethe

The love of color is primal and universal, no doubt built into our genes for survival purposes. The soul loves color because it appeals to a deeply held sense of beauty and inborn connection with Nature. Colors vibrate at various frequencies allowing you to feel the difference between warm red and cool blue on your hand or skin.

The garden offers up color as another mode of healing. If you are attracted to or suddenly desire a particular one, plant flowers in that color to help the healing process. You can also change your mood with the use of color.

The Taoist system of meditation offers some uses for color in healing. While meditating one can bring in the color of each organ and cleanse that organ using intention and thought. Red is found in the heart, white in the lungs, emerald green in the liver and gall bladder, yellow in the spleen, and dark blue in the kidneys. (I refer you to the many books by Master Mantak Chia.) In India, points on the body aligned with certain organs are called chakras. They are energy centers that relate to emotions and health and have colors associated with them. Meditations are available related to each chakra that aid in healing the organs and life issues.

You can use color to enhance the qualities you desire in your life. (Look into the Feng Shui chapters.) Learning to speak the language of color offers you additional tools to raise your levels of health,

"The Soul longs to be in places of beauty that inspire, feed our imagination, and create ideas. Beauty gives unlimited scope to the imagination and has the power to stir the soul."

Thomas Moore

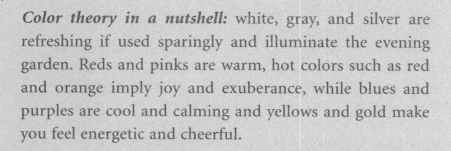

happiness, and success. When you see a color you react to it emotionally, physically, mentally, and spiritually. Color can boost your energy levels, soothe emotions, calm nerves, and stimulate circulation. It also can aid you in your quest for deeper knowledge of your inner self.

Our concepts of beauty come directly from Nature.

Red is used when we give the red carpet treatment to dignitaries. We say one is red-blooded, and we get caught in red tape. The negative side of red includes, red with envy, red-devil, and red-hot mad. Hummingbirds and butterflies flock to red flowers for their nectar. Many of our red dyes come from insects. Red leads to a hormonal response affecting the pituitary gland which then influences all other glands in the body.

Bringing red into your garden or landscape will increase the passion, movement, fire, love, vitality, and sexiness of your surroundings. Too much can increase irritability. I especially favor tulip, canna, geranium and rose.

Ruby, **wine**, and **scarlet** give a sense of power, majesty, and richness to any space. Use hollyhock, dahlia, and zinnia.

Pink used in the garden brings a feminine softness and gentleness to the area that also symbolizes love, innocence, and bouncy energy. We think of being tickled pink, looking at life through rose-colored glasses, and blushing pink. There are many shades of pink, including fuchsia and magenta. My favorites in the garden are monarda, cosmos, sweet pea, phlox, dianthus, and rose.

Orange is more radiant and actually hotter than red and will balance your personality, and lift your spirits. Many of the fast-food chains use orange in their logos because it stimulates the appetite.

When you incorporate orange in your garden or landscape you are bringing joy, earthiness, courage, and an adventurous spirit to your space. Planting gladiolus, marigold, and butterfly weed will infuse your garden and you with a sunny outlook and renewed interest in life.

Yellow stimulates nerve and brain tissue and if used in children's rooms will help to encourage good focus for studying. Yellow daisies in a hospital room bring healing, optimism, and cheerfulness. Yellow ribbons represent hope for the return of heroes.

As you use yellow in the garden or landscape be aware that it increases tolerance, self-confidence, clear thinking and enthusiasm, boosts the intellect, and builds optimism and harmony. I look forward to the blooms of coreopsis, hemerocallis daylily, and iris as mood lifters.

Green comes in more shades than any other color. We are surrounded by green. It is the most restful color to the eye. Perhaps that is why it is everywhere in Nature.

The green in gardens and landscapes brings to mind growth, life, renewal, rebirth, and calmness and will relax your breathing. I adore the many shades of green and especially how they are displayed in hostas, grasses, evergreens, and the variegated leaves of many plant varieties.

Blue has been used as a shield against evil forces for millennia. In the Orient, Middle East, and Africa, windows and doors are painted blue for protection against evil spirits. Blue is able to calm and relax people, although too much can lead to melancholy and depression. Many cultures use indigo dye because it is so readily available and easy to use.

Passion is the essential energy of the soul. It moves the soul to action; the result is emotion which is energy in motion.

33

As you walk through your garden or landscape notice how blue makes you feel. Many people assign the qualities of faith, devotion, constancy, and peacefulness to blue. Blue water makes you feel cool and refreshed and blue sky supports your dreams and visions. Any shade of blue (or purple) will bring colors that seem to clash into harmony with one another.

Delphinium, hydrangea, salvia, and lobelia are great additions to any garden.

Dark blue and **deep dark purple** symbolize insight, psychic ability, intuition, and introspection. When you wish to increase these abilities try using these colors in your environments and see if they accomplish the desired result. Iris, heliotrope, Johnny-jump up, violet, and pansy come to mind.

Purple is considered a royal and magical color that also increases one's ability to reach spiritual goals. Purple was the most costly dye known to the ancient world. It was harvested from a now-extinct shellfish off the eastern Mediterranean coast and was used exclusively by the wealthy, the nobility, and royalty.

Butterfly bush, iris, and larkspur are among my favorites.

Violet and **lavender** are recognized as soothing colors and aid in meditation, development of intuition and higher guidance, as well as the inner or spiritual quests of life.

I think of violet, iris, clematis, and bellflower as examples of this wide range of hues.

Salmon, **coral**, **peach**, and **apricot** are my favorite colors in the garden. They are actually a cross between orange, a little bit of yellow, and pink and combine the qualities of all these colors. I search nurseries and seed catalogs for any plant in this category. I like this color so much that I dyed my sheets and painted my bedroom in shades of it. I never get tired of these colors and sense they are working healing magic on me when I see them or come near them.

We feel surrounded by unconditional love, compassion, and acceptance that eases loneliness and soothes super-sensitivity if we use pink in the garden. Yellow builds self-confidence and optimism. Since orange can effectively unlock emotions, dispel anxiety, and increase self-esteem, placing plants of that color range in your garden or home will bring you their healing energy. When we use the family of colors containing salmon, coral, peach, and apricot, we are essentially using colors that combine yellow, pink and orange and are adding all their combined attributes and healing energies to our gardens and homes.

In my garden I use the azalea, canna, nasturtium, begonia, rose, daylilly, poppy, and foxglove to help satisfy my longing for this range of colors. Let me know if you find others.

Many thanks to Debbie Keller and her insightful book, *The Spiritual Garden,* for some ideas in the section on color above.

The following are not considered true colors, yet their use in the garden is acknowledged.

Brown makes us feel secure, stable, sheltered, comfortable, and homey because the earth is that color.

Black symbolizes mystery, formality, and seriousness. There are very few examples of true black in plants. Many choices come in chocolate, dark purple, and maroon that will achieve the same effect in the garden.

White in a garden can become a little boring and does not hold your attention for long, except at night when it lightens and brightens up a bed or planted area. It slows the mind to a point of stillness and quiet. Too much white can make one feel isolated or indifferent. Purity, sacredness, and wholesomeness are a few of the characteristics of white. In a garden or landscape it creates an aura of peacefulness and openness.

"A garden reflects a mood as surely as a fluttering fan or a blushing cheek."

Ralph Waldo Emerson

CHARTREUSE is a very special color in the garden.

Chartreuse enlivens a garden anytime, any season. This is one color that both men and women agree upon. It is a blend of green and yellow, goes with every other color in the garden, and serves as a spot and shade brightener. I use it as a contrasting color with dark green and as a softening remedy for the over-abundance of green in our landscapes.

SOME WELL-KNOWN CHARTREUSE PLANTS;

- Nandina
- Hosta "Sum and Substance," "August Moon," "Inniswood," "Little Sunspot," "June," or "Daybreak"
- Golden sweet potato vine
- Canna "Bengal Tiger"
- Creeping Jenny
- Ladies' mantle
- Euphorbia

BLACK, DARK PURPLE, MAROON AND CHOCOLATE:

Use black, dark purple, maroon and chocolate plants that are deep, rich, and almost black to bring an air of mystery to a garden. They go with every color and provide the eye with something new and fresh to enjoy.

For these selections go to www.johnnyseeds.com

- Tulip "Queen of the Night"
- Aeonium arboreum "Zwartkop"
- Chocolate Cosmos
- Viola "Springtime Black"
- Iris "Chrysographes"
- A nearly black hollyhock "The Watchman"

WAYS TO USE COLOR IN THE GARDEN

- Pick a single color and use flowers in all the tints, intensities, and hues of that color.

- Plant a bed of harmonizing colors such as yellow, orange, and red; purple, violet, and blue; yellow and blue; hot pink and apricot; pinks and blues.

- Blue or purple will make any combination of colors work well together.

- Too much red can make a garden appear smaller than it is.

- Create a contrasting color scheme from opposites or from one band of color next to another on the color wheel. (Color wheels can be purchased from artists supply stores.)

- Try a 3-1 ratio of pale hues to bright.

- For a bold look use groups of single colors, randomly repeating them, clustering at least three of the same plant in each group.

- Use the color green as you would a whole range of colors. It is the color in the garden that pulls everything together.

- Evergreens come in many shades of green: bright green, dark green, blue-green, teal, silvery green, gray-green, yellow-green, or chartreuse. They provide good examples of fine texture and contrast as well.

- Limiting your color palette by repeating a select few gives more impact and coherence to your design.

MORE ALMOST BLACK SELECTIONS FOR YOUR GARDEN

For these selections go to www.selectseeds.com

- Hollyhock "Nigra"
- Tobacco "Darkness", dark purple
- Morning glory "Grandpa Ott's", dark purple
- Cup and saucer vine, dark purple
- Cornflower "Black Gem"
- Castor oil bean "Carmencita", chocolate
- Sweet pea "Black Knight"
- Sweet scabious "Ace of Spades", dark maroon.
- Sweet potato "Blackie", dark maroon.

For these selections go to Dutch Gardens, 800-818-3861

- Redbeckia "Black Beauty"
- Gladiolus "Black Jack"
- Black elephant ears "Black Magic"
- Many types of asters and dahlias

37

Making the most of shape, size, and texture is just as important as using color in your garden. Create drama by using different textures and leaf shapes: long and short, pointy and round, wispy and dense, variegated and solid. Also consider fuzzy, velvety, smooth, waxy, furry, hairy, and shiny leaves as a form of texture. Look at bark and stem patterns as well.

Balance contrast and repetition:

Gardeners embrace gardens like old friends connecting at a reunion.

- Contrast creates excitement while repetition holds a design together.

- Bold and fine-textured species planted near one another provide visual interest.

- Repetition of certain elements moves the eye through the landscape.

- Study Nature and use plants as you observe them in natural settings.

How does color affect the soul? Color enhances our appreciation of the natural world and provides a deep source of pleasure. A black and white world may become boring. Color deepens the intensity of everything we see and experience. I have no desire to question that as I sit back and enjoy the scenery. That's soul enough for me.

🌿🌿🌿 *Garden Gift:* HERBED OIL 🌶🌶🌶

Place olive oil or vinegar in attractive bottles and add sprigs of red basil, green basil, rosemary, thyme, tarragon, or oregano. Cork and tie with raffia and give as gifts.

ROCKS

"When you tug at a single thing in nature you will find it attached to the rest of the world."
—John Muir

Rocks have many qualities that increase the soul of your garden. When you bring rocks into an indoor or outdoor space, you will feel the majestic yet humble, and energy enhancing attributes of these mellow beings. You will also feel connected to the timeless, divine, and spiritual aspects of Nature. Adam was formed of the earth, which is pulverized rock, and that is one reason you feel a natural kinship with rocks.

Both the Chinese and Japanese value mountains and therefore rock as symbols of the holy and transcendent in the natural world. "Stones are the bones of heaven and earth," says a Chinese philosopher. Rocks were featured in their gardens because they symbolized mountains, rock outcroppings, cliffs, riverbeds and ravines. The Chinese thought these natural forms contained hallmarks of the divine, the fingerprints of the gods. Mountains were viewed as centers and conductors of cosmic energy, and because they were thought to generate rain clouds, it was felt they were the source of the Empire's fertility. In Japan groupings of rocks in gardens and landscapes have subtle meanings related to Zen beliefs and practices.

Greek, Roman, and Celtic cultures believed that their Gods inhabited mountains and resided near peaks, under the mountains, or in caves. Judaism, Christianity, and Islam believe that mountains represent holy ground.

"Stones are the bones of heaven and earth."

Chinese philosopher

The Blarney Stone in Ireland, Stonehenge in England, prayer beads, gemstones on the breastplate of the high priests, jewels in crowns, and kissing the Pope's ring all represent the respect and status civilizations have given to stones of all kinds. Cairns or piles of stones mark burial sites, graves, and roadways.

Rocks and stones symbolize for us a feeling of solidity, stability, and security. They remind us of the timeless and eternal aspects of our connection with the Earth and serve as hallmarks of humility and unpretentiousness.

Rocks and stones have been used by every culture on the planet for ceremonies and marking important sites. Ancient cultures felt they housed divinity, had healing powers, grounded an area, and conducted energy.

One can also explore the use of crystals in garden beds or around a water feature. They add a quality that is indescribable as the sun catches their points and structure and creates a special glow to a space. A garden is rarely complete without rocks. After I built a water feature and edged it with rocks (mostly rounded ones), I felt better, as though the space was now grounded, balanced, and complete.

It is good to pick a single boulder or large beautifully shaped rock to become the guardian rock of your garden. It may be placed in a way that invites you to sit on it and meditate or view your garden. It serves to ground the area and provides guidance and protection for you and impacts your garden in positive ways. A smaller rock may be placed in an area to be used as an altar or a table for fairy gifts.

Rocks and stones come in all forms, sizes, and shapes, and create a sense of permanence and drama in the garden and landscape. When placed correctly, they accent and anchor plants and provide a great deal of visual variety and texture.

ROCK TYPES

Tall and Vertical: Remind us of austerity and determination. They tend to be taller than wide at the point where they enter the ground, have a vertical thrust, are used for waterfalls, and to simulate mountains and cliffs.

Large and Rounded: Imply steadiness, security, and authority.

Wider than tall: Evoke stony outcroppings.

Low and Flat: Stand for stability and placidity.

Arching: Represent activity and forcefulness, are vertical in nature, and are used when the eye needs to be pulled up.

Reclining or horizontal while being wider than high and sloping to one side with an irregular shape: Symbolize peacefulness and serenity.

Some final thoughts about rocks:
- Remember rocks are alive. They exist at a slower rate than the rest of creation around them.
- Treat them with reverence and care. They are beings too. Each stone has witnessed eons of events.
- Choose rocks for their particular qualities of shape, size, and coloration.
- Determine what is special about each rock that you are attracted to and use it appropriately.

Rocks contain the wisdom of the ages and may serve as anchors for your dreams and visions, while being steadfast protectors of your space. You may find yourself talking with your rocks, asking for advice, and sharing your innermost desires. All matter contains consciousness and has sentient qualities, including rocks, stones, crystals, and gems.

"We perceive Mother Nature as an 'it' not as a 'thou.'"

Philip Sutton Chard

41

NSECTS AND BUTTERFLIES

"Part of a deep sadness we carry with us as a species is the barely conscious loss of a loving relationship with the world around us."
—Claire Cooper Marcus

Try a new view of insects; see their beauty and reflect on their importance for the survival of plants and other beneficial life forms. Plants and insects developed and evolved at the same time eons ago.

Recently I watched a series of programs on HGTV, "The Secret World of Gardens," hosted by Martin Galloway, environmental biologist. This entertaining and educational show opened up a whole new world of wonder for me. I became aware of the unseen life that goes on under our noses in our gardens and began to understand how varied, beautiful, amazing, and complex insects are. I have new respect for them. Insects increase the soul of our gardens immeasurably with their indomitable spirit and resourcefulness.

Their casings and other remains enrich the soil and give it texture. Many insects clean up after other organisms that live in the soil. As they go about their work, they turn the soil and help to aerate it.

It has been reported that thousands of new species of insects are discovered every year. Referenced by The University of Arkansas, Etymology Department and the web at http://www.eduscapes.com/42explore/insects.htm.

"We are part of a web of interconnected life."

Pamela Jones

Insecticides eliminate beneficial insects. When we kill an insect, we inherit its work. The bothersome bugs don't disappear; they just go elsewhere. Our goal in the garden is to work with nature and her little annoyances. Life on our planet would cease without the interplay of insects of all kinds because they play a pivotal role in pollination of our crops.

Humor and psychological methods promoted by the manufacturers may lure you into buying various sprays and insecticides. They treat it as a war and tell you that only an industrial complex as organized as the military can take care of the job correctly for you. They tend to pit people against the whole world of insects; they are the foe and you are the victim. You weaken and then succumb to the advertising campaigns.

Maybe you have noticed that I don't use the word pest to name or describe insects or bugs anywhere in this chapter or elsewhere in the book. That is because I believe they are not pests but an integral part of Nature and I have decided not to use derogatory language describing them. It is my feeling this type of language relegates them to the category of enemy, removes us from any mutually beneficial relationship that we might have with them, and demeans them. We then feel superior and can decide to arrest their growth and control them. I have also decided not to use any war-like words such as attack, shoot, annihilate, kill, assault, fight, war, and combat. I try to maintain a positive regard toward beings that I admire and that benefit life.

SIX BEST BUGS FOR YOUR GARDEN:

Ladybugs

In North America alone, there are over 400 species of ladybugs. They require plant sap that has been partially digested in the bodies of aphids to survive. This might be the whole purpose of aphids. The ladybug is actually a species of beetle that feeds on aphids, mealy-bugs, white flies, and mites. They are attracted to yarrow and butterfly weed. Some experts recommend putting out a sugar solution around the edges of your garden or vegetable patch. This appears to encourage ladybugs to come to your area.

Beetles

This is another large category of helpful insects. They flit from flower to flower searching for nectar and pollinate many plants in the process. Beetles and ladybugs love pollen-bearing plants such as sunflowers, yarrow, sweet alyssum, clover, dill, tansy, and fennel. Refrain from using insecticides, as this will kill them. Before washing aphids off plants check you are not disturbing beetle larvae or the beetles themselves.

Praying Mantises

These insect-loving creatures prefer wildflowers and prey on more than 200 species of insects. Be careful when ordering from mail-order companies. They mainly carry mantises from China and Europe, and there is concern about these large and voracious immigrants.

Parasitic Wasps

These important bugs lay eggs in cabbage worms, loopers, and borers. When the infants hatch, they consume the host insect.

Green Lacewings

The hatching larvae eat aphids, mealy-bugs, mites, and white flies. They like cosmos, coreopsis, and angelica. You can order eggs from catalogs.

Bees

Bees are one of the most beneficial insects on the planet. They are the highly esteemed pollinators of most growing things and are very sensitive to insecticides and mites.

One school of thought says that a particular insect may have the potential to introduce a missing trace element to the soil and that it exists to restore an overall balance to the surrounding environment.

Spiders

"Certain insects are very important to the growth of rice and barley, specifically spiders."

Masnobu Fukuoka

These are voracious eaters of other insects and therefore benefit the garden.

One morning I woke up to find a large spider web glistening in the sunlight on my kitchen bay window where I scan my garden while eating. The intricate pattern this being had created was gone before the end of the day. Oh, but I had the pleasure for a few moments in time to appreciate how the dewdrops sparkled like diamonds on the ephemeral home of a kindred weaver. I am a weaver, too, and I will never forget that experience.

Small wasps

These wasps can only pass their larval stage when nourished by leaves digested by caterpillars or other insects. A colony of wasps is fed in its early stages on aphids. Maybe this is another reason for the existence of aphids?

Ordinary wasps

These are very helpful in the garden. They feed on other insects and the remains of other insects, especially caterpillars that can destroy crops.

Human action of some kind or Nature herself may initiate an imbalance or create weather conditions which are conducive to the rapid proliferation of one or another insect species. A number of hard frosts in February and March will delay the emergence of aphids from hibernation. The accompanying sunshine may tempt the ladybugs to come out. Finding no nourishment, they perish, and when the aphids emerge there is nothing to control them. It may take some time for the predator population to build up and take care of the imbalance that has occurred. When this situation occurs gardeners and farmers are generally faced with a dilemma; let Nature take care of the imbalance or intervene in some way with the process so crops and plants will not be lost. I advocate using organic and nontoxic means of controls. The easy way out is not always the best for the overall health of the environment.

SPRAYS TO ALLEVIATE AN INSECT PROBLEM

Mix these in plastic spray bottles and use on your garden or houseplants. This is just a sampling of the many available which can be used as repellents or insecticides.

Nettle spray for aphids and caterpillars:

(Nettles can be used as a foliar spray and tonic for many plants.) Fill a container with nettles gathered at any stage of flowering or buy nettles at the herb store that are either dried or fresh. Add water until they are immersed. After 24 to 48 hours, strain the liquid and dilute to 1 part nettle juice to 4 parts of water.

A general insecticide:

1 cup chopped hot red pepper (I have read that 1 tablespoon of cayenne will work too)
2 large cloves of chopped garlic
1 quart water
Steep for 24 hours, then strain and use.

Murphy's Oil Spray:

Add 1 teaspoon Murphy's Oil Soap to 1 quart of water. I have used this successfully on a small outbreak of aphids that appeared on a climbing rose.

Rose spray:

4 tsp. baking soda
4 tsp. cooking oil
2 tsp. liquid dish soap
Mix all the above ingredients with 1 gallon of water and spray on rose bushes to eliminate aphids and other insects.

"When insecticides are used, spiders are decimated and cannot do their work of controlling the leaf hopper population."

Masnobu Fukuoka

47

For aphids and other insects:

Add 1 tablespoon cooking oil plus 1 tablespoon dish soap to 1 quart water. Mix well. Spray on the affected plants.

GARLIC BUG-BE-OFF:

This spray has been known to eliminate and prevent many types of bugs. Take 2-3 tablespoons of garlic powder and 2 tablespoons of Tabasco sauce; add a dash of mild dish soap for each gallon of water. Spray the infected plants on a 5-7 day interval until control has been achieved.

Select Seeds has three products that help with insect problems: Neem oil soap, hot pepper wax insect repellent, and aphid alarm. Call 860-684-0310 to order a catalog or go online to www.selectseeds.com. Also try Gardner's Supply Company call 800-427-3363 or go to www.gardeners.com.

COMPANION PLANTS

Used by farmers and gardeners for centuries, companion plants help to control an insect that is creating a problem for a particular vegetable, tree, or herb. The following plants can help control the insect population organically to help keep the ecological balance in your yard and garden. The most common plants to try are chives, coriander, pennyroyal, sage, onions, thyme, wormwood, radish, calendula, mums, scented marigold, nasturtium, tansy, yarrow, and basil.

- Garlic aids in the prevention of many bugs. Try a clove of garlic around the roots as you plant roses to prevent aphids and other insects.

- Planting parsley near roses increases their fragrance. Yarrow is said to help most vegetables and herbs increase their aroma. All vegetables are aided in some way by planting herbs nearby.

I have read that insects actually protect plants by seeking out the weak ones and destroying them to ensure the genetic strength of the entire plant species.

• Plant nasturtiums, yarrow, marigolds, tansy or radish near herbs and tomatoes to ward off insects.

• Pennyroyal, mint, or tansy will help an ant problem.

PLANTS THAT PROVIDE PROTECTION FROM INSECTS

PLANT TO PROTECT	GUARD PLANT	INSECT REPELLED
asparagus broccoli, cabbage	basil, calendula, parsley sage, thyme	asparagus beetle cabbage moth
cauliflower	onions	rust flies
carrots	rosemary parsley	carrot flies carrot beetles
corn	scented marigold	various insects
cucumbers	radish, nasturtium	cucumber beetles
green beans	scented marigold	bean beetles
potatoes	horseradish	beetles
squash	radish, nasturtium, tansy	squash bugs
tomatoes	basil scented marigold tansy	hornworms various insects flea beetles
roses	scented marigold chives garlic tansy	nematodes aphids ants Japanese beetles
apple trees	nasturtiums	aphids
plum trees	horseradish	ants

The **Three Sisters** is one of the most efficient biological insect suppression methods used by Native Americans, who taught it to early settlers. Three vegetables are grown in unison. Plant corn, encourage beans to grow up on the stalks, and plant squash to trail on the ground. This method suppresses weeds and insects and uses the land to its maximum.

Botanicals that are totally natural can also be used for pest suppression. Neem is extracted from the neem tree in India and has been used for centuries. It is believed that insects will not develop a natural resistance to it. Rotenone and pyrethrums (from a mum) also work to check insects and are very powerful. Use them with care and follow the instructions carefully. Check Resources for websites related to insect control.

Refugia is a new method of plant interaction that has drawn attention in recent years. It endorses using plants that provide an environment especially beneficial to parasitic insects, thereby keeping the rest of the insect population in check.

Beneficial nematodes are used to control larvae of termites, ants, fleas, beetles, cockroaches, and to eliminate many other garden bugs, such as corn borers, cutworms, and cucumber beetles.

PROOF THAT NATURE CAN TAKE CARE OF ITSELF:

A bug ate the leaves of my beautiful red hibiscus until they all looked like lace. I was horrified. I asked Carl Totemeier, a professional horticulturist, who said the plant would recover if I did nothing. I listened, did nothing and waited. Patience is a way of connecting to one's soul. After a while the plant sent out many new leaves even though most of the original ones still resembled lace, and it produced many beautiful flowers despite its unhealthy look. This year it grew bigger and taller and bloomed until October. Just as we strengthen our immune systems by weathering colds, flu, and the like, plants also seem to get stronger after dealing with a health issue.

ESTABLISHING A BUTTERFLY GARDEN

The most successful butterfly garden includes plants that meet the needs of butterflies during all four stages of their life cycle: egg, caterpillar, chrysalis, and adult.

After mating, the female butterfly searches for the best plant to lay her eggs. Monarchs lay on milkweed, swallowtails on parsley, tulip tree, and wild cherry; others use the passion flower. In a few days, the caterpillars emerge and eat only one kind of plant; the one on which the female laid her eggs to ensure survival. The caterpillars mature in a few weeks, shed their skin several times and change into a chrysalis from which the adult butterfly emerges. After hatching from the chrysalis the adult butterfly looks for nectar-rich flowers to feed on, and the whole process begins again.

SOME OF THE MANY PLANTS FOR A BUTTERFLY GARDEN
Nectar Sources

- **Shrubs:** Azalea, butterfly bush, glossy abelia, lilac, and spicebush.

- **Perennials:** New England aster, pineapple sage, purple cone-flower, sedum, thrift, vervain, phlox, butterfly weed, monarda, milkweed, violet, scabiosa, yarrow and Queen Anne's lace.

- **Annuals:** Cosmos, French marigold, heliotrope, impatiens, Mexican sunflower, verbena, zinnia, lantana, and penta.

HOW TO CREATE A BUTTERFLY GARDEN

1. Find a sunny spot.

2. Plant nectar-producing flowers.

3. Select single rather than double flowers

Butterflies are an ancient symbol for the soul and for transformation.

PLANTS BUTTERFLIES
USE FOR PROPAGATION

Host Plants	Butterfly attracted
Butterfly weed, milkweed	Monarch
Passionflower, violet	Fritilary
Spicebush	Spicebush swallowtail
Tulip tree, apple, cherry, ash	Tiger swallowtail
Pawpaw	Zebra swallowtail
Parsley, dill, fennel, rue	Black swallowtail

(the double ones are harder for them to get into and light on).

4. Use large splashes of color. Butterflies are attracted to flowers by their color and prefer reds, oranges, purples, and bright pinks.

5. Plan for continuous bloom throughout the season and into the fall.

6. Include host plants that provide food for caterpillars and lure females into the garden to lay their eggs.

7. Include damp areas, shallow puddles with mud (they derive needed minerals from the soil) and a water feature to provide them with water to drink and to clean themselves.

8. Place flat stones around to encourage perching while allowing them to spread their wings and bask in the sun.

9. Use biological controls for insects. Insecticides and herbicides are toxic to butterflies. Be wary of Bacillus thuringiensis which is used to kill the codling and cabbage moth. It is toxic to butterflies.

10. Become a butterfly watcher and purchase a field guide.

11. Butterflies are around from 10 a.m. to 3 p.m. Because that is the hottest time of the day, it is wise to prepare a cool or indoor location to view them.

12. Try to use drip or ground-level irrigation methods because overhead sprays wash away the nectar they need.

For more information contact:
www.msue.msu.edu/genesee/natres/buttfly.htm

I think you can tell that I love all of Nature, even bugs. They are already here, so why not get along with them? Oh, it's easy to love butterflies; they are so beautiful and don't bother anyone or anything. It's those pesky bugs that drive everyone to distraction. They are here to stay and must have some purpose. I love insects—always have and always will. I think they are filled with soul because they reflect the wonder and diversity of creation. They just "are" like soul just "is," and that's good enough for me.

> *🍃 🍃 🍃 Garden Gift:* **AROMATHERAPY** *🍃 🍃 🍃*
>
> Bathe with 7 to 15 drops of a mixture of 2 to 5 fragrant essential oils dispersed first at the tap.
>
> **Relaxing**: Lavender, thyme, lemon verbena, comfrey or chamomile.
> **Stimulating**: Jasmine, peppermint, yarrow, or ginger.
> **Soothing**: Rose, sage, rosemary, or calendula.

"Those people who I most admire are those who have turned away from the cluttered materialism and returned to the beauty of the natural world and objects made by human hands."

Joanna Macy

BIRDS AND WILDLIFE

"I find you Lord in all things and in all my fellow creatures, pulsating with your life."
—Rilke

Birds bring activity and exuberance to a garden, making our spaces feel alive, cheerful, and full of optimism and hope. It is thought their song as well as the chirping of cicadas encourages tree growth.

SOME IMPORTANT CONSIDERATIONS REGARDING FEEDERS

- Keep your feeder(s) well maintained because birds can carry fungus, viruses, and bacteria from one place to another and infect each other.
- Clean feeders with a diluted white vinegar solution or hydrogen peroxide every few months. Mix 1 part vinegar to 10 parts water and rinse thoroughly. Refrain from using any household cleansers. These contain chemicals harmful to birds.
- Do not fill a feeder with moldy, smelly, or dusty seed, or offer birds moldy bread.
- Make sure your feeders have no sharp edges.
- Try putting mulch around the base of your feeder. The lack of plant cover makes it difficult for a cat to prey on the birds and surprise them while they are busy feeding.

"A genius is the man who has the eye to see nature, the heart to feel nature, and the boldness to follow nature."

Frank Lloyd Wright

Scientists are creating a new technology that is taking its cue from birds. By mimicking a flock of birds in formation, scientists are discovering that planes can imitate this pattern and save 12 percent in fuel, adding up to about 1 million gallons a year. This way of flying, it was discovered, will avoid drag and increase lift.

"NBC News," Sunday, December 9, 2001

• It is best to have a variety of feeders, keeping the amount of birds at each feeder down so they are not crowded.

• You may find that the attractive blue jay is a noisy bird that menaces and bothers other smaller birds, chasing them away from the feeders.

WAYS TO ATTRACT BIRDS

Planting types of flowers that produce a multitude of seeds such as asters, coreopsis, agastache and sunflowers will attract goldfinches and songbirds. Leaving seeds, pods, and berries on plants for all types of birds to eat in the winter months is good. Providing a variety of bushes such as elderberry and high bush blueberry, for them to perch on and which also offer shelter, protection, food, and nest-building opportunities is also beneficial. It is important to provide a source of water and a variety of feeders to attract all kinds of birds. Allow rotted limbs and trunks of trees to remain on your property for birds to feed on the insects inside and to use the debris to build nests.

Crows can be controlled by not offering them cracked corn. However, they have their place in nature and are considered by the Native American to be protectors of many birds.

Caution:

Don't paint the insides or outsides, or stain the birdhouse, as this can harm the birds. Metal roofs can create too much heat and injure birds. Bird lovers avoid using insecticides and herbicides in their gardens because they have a toxic effect on the birds and the insects they eat.

BIRDWATCHING

Start to identify the birds that frequent your yard and your feeders. This will increase your pleasure and provide a wonderful educational tool for your children. Bird enthusiasts spend $83 million a year bird feeding, making it one of highest hobby expenditures in the United States today after golfing, fishing, and gardening. For more information try www.birding.com and www.audubon.org.

- Arrange a place to feed the birds near where you eat or watch TV.
- Watch the birds as you start your day, sipping coffee, or reading the paper. It is a form of meditation and calms the spirit.
- Trees and bushes provide good protection and nesting sites and will attract many varieties of birds if they produce berries later in the season.
- Clean birdhouses after each nesting to eliminate parasites.
- Birds won't nest close to one another, so space birdhouses some distance apart.

WILDLIFE

Animals in the garden deepen our appreciation of Nature. Observing their curious habits, life concerns, and tender ways, enhances our ability to connect with our own soul.

TURTLES

I offer for you a spiritual or indigenous wisdom approach to understanding turtles.

"The turtle represents wisdom and teaches patience. It follows the path of working in Truth. The turtle is able to carry the weight of many possibilities to any given situation, is not in a hurry to make

"He who shares the joy in what he's grown spreads joy abroad and doubles his own."

Unknown author

57

quick decisions, and has a strong heart and a keen mind that work together as he/she considers carefully how all life will be affected by his/her decisions." Taken from *The Wheel of Life* by Howard Isaac.

SQUIRRELS

Wildwood Farms (see Resources), where "squirrel feeding is our specialty," recommends feeding squirrels on the opposite end of the property from birds so they will leave the bird feeders alone. Research has shown they are very territorial and when fed on a continuous basis, away from bird feeders, they mark their area and keep other squirrels from entering the area, thereby reducing the squirrel population to one family that will leave the bird feeders and garden alone.

DUCKS AND GEESE

"When a family of ducklings fell down a Vancouver sewer grate their mother did what any parent would do. She got help from a passing police officer. Vancouver police officer Ray Peterson admitted he was not sure what to make of the duck that grabbed him by the pant leg while he was on foot patrol on Wednesday evening in a neighborhood near the city's downtown. 'I thought it was a bit goofy, so I shoved it away,' Peterson told the 'Vancouver Sun' newspaper. The mother duck persisted, grabbing Peterson's leg again when he tried to leave, and then waddling to a nearby sewer grate where she sat down and waited for him to follow and investigate. 'I went up to where the duck was lying and saw eight babies in the water below,' he said.

Police said they removed the heavy metal grate with the help of a tow truck and used a vegetable strainer to lift the ducklings to safety. Mother and offspring then departed for a nearby pond." —Taken from Reuters News Service on July 13, 2001

Birds and wildlife are part of the natural world around us. They enrich every day for me as I watch and take part in their lives. There is no question in my mind that they bring soulful qualities to me and to anyone's garden. Each spring I look forward to the return of the daily morning birdsong greeting. Consider for a moment, a world without birdsong and other wildlife sounds.

If you like the duckling story, you will also like the movie "Fly Away Home" with Jeff Daniels.

LESSONS FROM GEESE, *Author unknown, (but inspired by the Divine in nature).*

As each goose flaps its wings, it creates an
uplift for the bird following. By flying in a V
formation, the whole flock adds 71% more
flying range than if each bird flew alone.
**Lesson: People who share a common direction
and sense of community can go farther and
get where they are going quicker and easier
because they are traveling on the thrust of one another.**
Whenever a goose falls out of formation, it
suddenly feels the drag and resistance of trying to
fly alone, and quickly gets back into formation to
take advantage of the "lifting power" of the bird
immediately in front. **Lesson: If we have as
much sense as a goose, we will stay in
formation with those who are headed
where we want to go.**
When the lead goose gets tired, it
rotates back into formation and
another goose flies at the point position.
**Lesson: It pays to take turns doing the hard
tasks, and sharing leadership with people who,
as with geese, are interdependent with each other.**
The geese in formation honk from behind to encourage
those up front to keep up their speed.
**Lesson: We need to make sure our honking
from behind is encouraging, not something less than helpful.**
When a goose gets wounded or sick or shot down,
two geese drop out of formation and follow it down to help
and protect him. They stay with the goose until it is either able
to fly again or dies. Then they launch out on their own with
another formation or catch up with the flock.
**Lesson: If we have as much sense as geese,
we'll stand by each other like that.**

EEDS

"The ecological disintegration that is occurring in the environment around us is echoed in the inner landscape of the soul."

—Denise Linn

Weeds are important in the balance of nature. If it were not for the constructive work of several weeds, the last wild-grass pasture/prairies in this country would be as dry and barren as a desert. A field of weeds can survive drought conditions. Weeds help to stabilize the soil, prepare it to maintain all kinds of grasses, and work to unlock compacted, dry earth. It is thought that weeds such as thistle and milkweed help native grasses survive erosion and dust storms and help soils retain moisture. They are among the most ancient plants; ragweed and crabgrass grew on the planet 4,000 years ago. Most weeds have either a medicinal or nutritional value; some are just plain beautiful.

Grasses are the largest family of flowering plants on the planet and all our grains are essentially grasses. They began as weeds that were cultivated for human use. Left alone to take over fields, some would call them weeds. There are 2,000 named species; many others have not been categorized yet. Wild plants, flowers, and herbs are weeds, too.

We classify them as useless and treat them as though they were on the FBI's most wanted list. "Weeding can attain the status of a holy war," says Diane Ackerman, in *Gardening Delight*. I admit that weeds

"Absence of a bond with the earth erodes our interpersonal bonds with each other."

Philip Sutton Chard

Peter Gail may be the world's foremost authority on dandelions. He is the founder of Defenders of Dandelions, president of Goosefoot Acres Center for Resourceful Living, and publisher of his books, *The Dandelion Celebration* and *The Great Dandelion Cookbook.* He also runs the National Dandelion Cookoff held in Dover, Ohio, which attracts up to 20,000 people. To reach him call 800-697-4858 or e-mail petergail@aol.com.

From *Country Living*, May 1998, "In Defense of Dandelions" by Maggy Howe, pages 112 & 116. (See Resources for more references)

are annoying because they seem to destroy the balance and beauty in our gardens. Nature in its wisdom has reasons for their existence. According to Steve Brill in *The Wild Vegetarian Cookbook*, people don't realize that many of the plants they try to eliminate from their gardens are not only edible, but are delicious if prepared properly. I believe if we developed a more soulful approach to our environment we would find uses for all weeds.

I often succumb to the weed-pulling urge too. In fact there were a lot of weeds around one of my apple trees and I decided to pull them up. In the process I relaxed into the job and looked at every weed with its healthy deep-set root system and began to admire them instead. I began to like this chore as it became therapeutic, relaxing, and meditative; it emptied and focused my mind. I noticed the soft and aerated soil around the tree; the weeds had accomplished that. All looked good again as I covered the base of the tree with a new dose of mulch. I had a new respect for weeds as I dug them out and put them on the compost pile, knowing their lives went on enriching other soils and plants.

It's true that weeds in the garden take light, moisture, and nutrition from our plants, and their very existence signifies deeper problems going on in the soil. They warn of compaction of the soil, lack of aeration, loss of humus, and a decrease in fertility levels. They grow and multiply when these conditions are present, reminding us to replenish and renew our soil. Even in a well-tended garden, they seem to appear from out of nowhere. Remember that many weeds are borne by the wind or carried by birds to your space; some appear after you till the soil. Relax, step back, acknowledge their purpose, and then perhaps, try some of the remedies I offer in this chapter.

DANDELIONS

A weed is just a plant that is out of place.

Dandelions are a special category of weed; practically everyone has them. In Italy they are prized for making salads and wine in the spring. These special weeds are excellent for cleansing and enriching the blood after a long winter of not having many fresh vegetables or greens to eat. They are also considered an excellent liver tonic and digestive aid. Europeans have understood the real value of this much-maligned weed. It was used as a spring tonic all over Europe for thousands of years. Many people are raising them in their herb and greens beds now and are raving about the fresh tart taste, somewhat like chicory. Cleaning and drying their roots to make a winter tonic tea is also a good use for them. (From Karen Kreiger.)

Many sources say they are rich in nutrients and very tasty when harvested in the early spring before their leaves get too tough and bitter. Some sources suggest snipping them from your lawn before you mow. Be careful, they haven't been visited by dogs first. Clean them by soaking and rinsing several times and then sauté as you would kale and spinach, or add to salads.

Native to every continent, dandelions have more vitamins A and C, iron, calcium, magnesium, and potassium than broccoli, carrots, or spinach. They remove excess water from the body without depleting potassium, lower cholesterol, stimulate saliva and bile, and help with indigestion and constipation. Weight loss, improved functioning of kidneys, pancreas, spleen, and stomach have also

63

been cited. You can get less bitter varieties from seed catalogs. They were used for food until the 1960s when lawn-care companies dubbed them as a weed and launched advertising campaigns that overlooked their nutritive value and advocated eliminating them from all lawns.

OTHER WEEDS

Many plants that we value in the garden began as weeds such as yarrow, morning glory, comfrey, larkspur, and poppy. Yarrow is very beneficial in the garden and to butterflies. Some weeds have medicinal value, such as burdock root, comfrey, calendula, and pokeroot. Lambs quarters are very good in salads. Teasel has been used by weavers for centuries to brush the nap on fabric.

WAYS TO WORK WITH WEEDS

1. Weeds are specialists; choke them out by planting faster, thicker growing crops.

2. Pick plants that naturally repel certain weeds.

3. Add rotted manure, lime, or compost to the soil to enrich it, discouraging most weeds from taking root.

4. Bring in earthworms to eliminate compaction.

5. Use green manures such as alfalfa, barley, oats, buckwheat, rye, and soybeans as a cover crop. Plow them under to enrich the soil.

6. Spreading wheat or other types of straw or mulch in your garden will also keep weeds down. This decreases the amount of air and sun available for growth. When using straw it is a good idea to make sure that it is free of seeds, including weed seeds.

7. Cut down annual weeds before they set seed because they produce millions of them.

The carrot came from the wild, as a weed. It started out as Queen Anne's lace.

64

8. Mow high, about 3 inches or so, to discourage weeds from growing due to the shade that higher grass affords.

9. Perennials such as dandelions, wild garlic, and curly dock can be pulled after a rain or soaking. Of course save the dandelions for salads and wine making.

10. Doing the last four tips periodically is a lot safer, cheaper, and healthier than using chemicals.

The topic of noxious weeds is out of the scope of this book. It is a good idea to check with your County Extension Service and get help in identifying which ones appear in your area. Also, find out if the county is planning to spray chemicals on the road and right of way near your property. You can then inform them that you will take care of it yourself with non-toxic solutions.

The Nature Conservancy can give you more help with weeds: tncweeds@ucdavis.com. Also look into *The Gardener's Weed Book: Earth-Safe Controls* by Barbara Pleasant. Try corn gluten as an alternative to herbicides for lawns. Nitron Industries stocks it and gives directions for its use. Try www.nitron.com as well as www.attra.org for more information.

Have patience with your weeds. They might just be showing you the way to a healthier, more productive, and soul-enriching garden experience. Look at weeds and visualize them as food, flowers, or medicine. Thank them instead of being irritated with them.

"What is a weed? A plant whose virtues have not yet been discovered."

Ralph Waldo Emerson

🌿 🌿 *Garden Gift* 🌿 🌿

Fresh herbs from the garden can be put in a dish in the back of the refrigerator to dry out. Store in jars or condiment containers when done. I find this method better than drying them upside down for a week. It seems to preserve their flavor and color better and is akin to freeze-drying. You can also put several pinches in an ice cube and freeze them for later use in soups and stews.

PART 2

CULTIVATING A GARDEN OF DELIGHT

" . . . garden of dreams, garden on the
oasis of a life-drenched planet, garden
where desire finds form, garden of floral
architecture and speckled fawns, garden
where wonder is incised on a pebble
millions of years old, garden visibly and
invisibly teeming . . . garden that's an urn
for the soul."

—Diane Ackerman

HUMOR AND WHIMSY IN THE GARDEN

"When we learn to see such common objects as stoves and bath-tubs as symbols of nourishment and transformation, the rigid boundaries that tie our hearts and minds can become transparent to the workings of the spirit."

—Anthony Lawlor

Since a garden is a blank canvas, gardeners often spend their creative moments thinking up funny, whimsical, and downright rib-scratching enticements. There are many ways to have fun and bring humor into our lives. Lately, I have observed how gardeners are stretching their imaginations by venturing into the area of garden humor and whimsy. Planting in old chairs, wheelbarrows, and broken pots is one way. Another is, placing the unexpected on a path or in a flowerbed.

Gardeners use their imaginations when they find ways to make us smile as they create new lives for practical and overlooked objects. A gardener here in Fayetteville, Arkansas placed her mother's old mailbox in her garden and filled it with garden tools to avoid going to the garage for everything. This same gardener employed a painter to use a faux-painting technique that transformed her cement patio into beautifully colored flagstone.

I recently decided to check out the following web site: Garden

"The earth laughs in flowers."

Ralph Waldo Emerson

69

Humor, http://home.golden.net/~dhobson/constrange.html and found the following on June 18, 2002. "I manage a garden center, and I'm not sure if my customers or employees are the strangest thing there! Maybe customers who describe a plant as having green leaves! Cherie." I'm going to visit it every so often, when I can use a chuckle.

I observed the insects that appeared on the HGTV program "The Secret World of Gardens." Their dance-like movements, big eyes, astonishingly bright and beautiful color patterning, and many flexing appendages are at once funny and delightful. When we don't feel threatened by them, we can appreciate the delicacy of their form and the intense activity and liveliness they bring to our gardens. They rush about looking for food, mating, laying eggs, and succeeding in a dangerous world; we should admire them.

Squirrels are also humorous. I watch them with my dog through my kitchen bay window as I eat, read the paper, and do various kinds of correspondence. They chase each other all over the yard, engage in a constant effort to outwit the birds at the feeders and literally fly through the air jumping from one tree branch to another. I also love their curious nature.

This past spring, I saw a house finch grab a hanging branch on my silver lace vine. I watched in amazement as she turned over and over in mid-air, tugging and twisting a branch that she wanted for her nest. She let go for a while and then came back to try again until she had her prize. She entranced me with a truly acrobatic performance.

I watch birds play in my little pond, splashing, bobbing, washing, drinking, and having fun. You can hear their song of life as they scold, warn, encourage, and support one another.

Sometimes birds touch your soul in ways that not only make you laugh but tickle that spot inside that yearns for wholesome goodness and lightheartedness. This past spring a sparrow built her nest right in the center of the wreath on my front door. You can imagine my surprise and joy when I discovered five little blue eggs in there and then watched as they hatched and matured from fuzzy little beaked beings into full-fledged birds. I took pictures of their growth and felt they were part of my

Humor is very good for the soul, because laughter encourages the soul to look at itself lightheartedly.

family. That experience made me chuckle every day as I opened my front door in the morning to get my newspaper and felt a bird whiz by me into the garden. She had been on the nest all night and came back each evening to watch over her progeny.

This spring I was amazed at the antics of a wren outside my kitchen window; where I had placed some birdhouses on a table. She landed on one house, flapping her little wings and singing the most beautiful song. She did this continuously over several days and also spent some time looking up at me and singing, flapping her wings, and then flying up to a post nearby. She obviously wanted my attention. Finally, I figured out what she wanted. I hung the birdhouse on the post. She continued her behavior for several more days, began to build a nest in the house, and then attracted a male and now they are raising a family. Her ability to catch my attention, communicate with me, and finally succeed in her quest amused me and captivated my soul for days.

Frogs make me smile too. I have a tree frog that lives near my pond in the backyard. He trills his musical outpourings at times when he is looking for a mate. I have thought to myself; how can he hope to attract a woman in my neighborhood? Yet he does and tadpoles appear in my pond; then, he begins all over again. There is something elemental, basic, and sensual about his yearning, and the fact that he shares it causes visitors to my home to erupt into laughter and sport sheepish grins.

I have found handmade pieces that make me smile and placed them in my flower garden. As I do a walking tour of my flower beds, deadheading, picking, and trimming, I come upon a whimsical frog jumping up to greet me and a birdhouse embellished with bug sculptures, that says 'welcome' on the front door. I walk by a garden bird constructed of discarded old parts of farm and garden tools. He has two eyes made of scissors turned upside down and is absolutely hilarious. They make walking in my garden a veritable standup comedy act. Having such boundless child-like fun is good for your soul and certainly for your well-being. It is often said that doctors advise patients to incorporate laughter into their lives for its healing benefits. The garden presents many such opportunities.

Creative gardeners have turned old sinks and faucets into water features. Even old toilet seats have been turned into planters; the sky's the limit. Creating fanciful creatures from old rusty things you have around the house is a wonderful way to use your creative talents, place whimsy at arm's reach, and recycle at the same time. These humorous creations can start a fit of laughter that brings tears to one's eyes.

Garden art honors and celebrates life and the creative spirit as it stimulates our inner world. The special qualities of the art expresses the uniqueness of the gardener, enhances the drama of the landscape, and fills in gaps that nothing else could. I am sure you have humorous and whimsical garden stories of your own. I never feel downhearted in my garden. I am treated to flowers blooming with cheerful, carefree abandon, indescribable perfumes, laughter from all forms of art, and live performances of humor and whimsy. These give my soul nourishment as much as prayer would.

$\mathcal{T}$HE SENSES

"Man has no body distinction from his soul; for that called Body is a portion of Soul discerned by the five senses, the chief inlets of Soul."
—William Blake

Our emotions, feelings, and senses allow us to feel alive and part of Nature. In *The Healing Earth*, Philip Sutton Chard tells us that emotions arise from nature and then manifest in humans. He says, "Feelings are not exclusive to people, but rather reflect our participation in the larger emotional life of the planet."

We experience the emotional expressions of Nature through our senses:
- We say the wind is wild, soft, and gentle.
- Clouds can be ominous, dark, and brooding.
- Storms rage and fume with anger and fury.

Native Americans understood this phenomenon very well when they gave names to people, places, events, and animals that described their elemental nature and characteristics. Indigenous people experience Nature as their extended body in a world composed of multiple intelligences. Our culture is one of the first to lose its "ancestral reciprocity with the animate earth," says David Abram in *The Spell of the Sensuous*.

The senses are the primary way the Earth can inform our thoughts and guide our actions. It is only when we have direct

"One by one our senses are captivated and charmed by the garden."

Jeff Cox

73

sensory interactions with the land around us that we can appropriately respond to the needs of the living world. Senses have served all species on the planet well over the millennia, aiding survival. "The Earth tells us about ourselves and that she cares for us as a mother does, unfailingly," says Jeff Cox in *Creating a Garden for the Senses.*

Everywhere around you the natural world lives and breathes. The Earth speaks to us through her fragrances, sounds, textures, and other sensual and dazzling reminders of how passionate and rich our experiences can be when we connect with her.

TOUCH

The sense of touch is a direct extension of our skin, which is our largest organ. We can touch something and know if it is heavy, soft, light, metal, wood, velvet, silk, liquid, or solid. Touch teaches us that life has depth and contour and makes our world a three-dimensional one.

- As we touch plants, trees, and flowers in our gardens, memories from our childhood can flood our consciousness.

- Touch causes chemical reactions and physical changes that promote growth and pleasure.

- Touch is the first sense a baby uses to get to know the world and make sense out of his/her experiences.

- Children not held enough will have stunted growth and development.

- Heart rate, metabolic rate, temperature, brain-wave patterns and immunity can all be influenced by touch.

- Scientists tell us that petting and stroking animals lowers blood pressure and reduces stress.

A breeze tickles our skin as it wafts through the garden. Rain falling on us feels like soft caresses. Leaves come in textures that feel silky, leathery, suede-like, hairy, or smooth. We are enticed by the soft, velvety texture of a rose petal and put off by the sharp, prickly spines of a cactus.

Many tactile experiences occur in the garden: from walking barefoot on the grass to feeling the shaggy bark of a river birch, the velvety leaves on a mullein or lamb's ears, and thorns on rose and berry bushes. We touch with our eyes, as we take in the whole range of tactile offerings and know what each one feels like without actually reaching out for it.

SOUND

- Sound can mend bones.

- Music has soothing, relaxing, and healing qualities.

- Sound and prayer can sprout seeds held in the palm.

- Cows become more relaxed and give richer and more milk when Mozart is played in the barn.

- Plants are known to grow faster and hardier, and contain more nutrition when music is played near them.

- Our endorphin levels rise and our pupils dilate when we sing.

The buzzing of flies, the sounds of frogs croaking and birds chirping, the rustling of leaves, and the other sounds of Nature's chorus once were our symphonic backdrop. The sound of the earth under our feet, crackling leaves in the fall, or the repetitive sound of a hoe slicing through the earth mesmerize. Now environmental sounds from machines and appliances drown out the natural sounds.

"We are surrounded by the rhythm of Nature."

Christopher and Tricia McDowell

75

The Earth offers up sound in a myriad of ways for our benefit and healing.

- All the variances of rhythm and sound are played out in Nature's orchestra.
- Winds screech and roar and speak to us of gentleness and yearning.
- Waters laugh and call to us with their tempo and rhythm.
- Insects buzz and whirl as they go about their frenetic life activity.
- Crickets and grasshoppers chirp in tenor, alto, and bass voices.
- Birds sing to announce their presence, mark their territory, impress a mate, and boost their status.
- "Frogs get together in wet places and sing about sex and desire," says Diane Ackerman in, *The Natural History of the Senses*. I lived in North Carolina and will never forget the symphonic virtuosos of tree frogs all June and July. It sounded like Bach or Beethoven to me.
- Each tree has a distinctive voice when the wind rustles through its branches.
- Grasses and bamboo also have unique sounds as the wind strums their stalks like a guitar.

Silence is a source of healing that Nature provides as a free gift when we visit forests, mountains, caves, and isolated areas. We are so accustomed to man-made sounds that we get distracted from our own inner voice and forget to listen to it.

SCENT

- Smells go immediately to the limbic portion of the brain and are associated with emotion and memory.
- Scents stimulate learning, retention, relaxation, and our evaluation of things and warn of danger.

- "A flower's fragrance announces to the whole world that it is fertile, available, and desirable and its sex organs begin to ooze nectar for its suitors," observes Diane Ackerman. We sniff their aromas and are reminded of optimism, passion, youth, vigor, and joy.

- Humans as well as animals give off pheromones that attract the opposite sex. They are powerful indicators of availability and readiness to mate.

- Plant aromas are meant to attract bees, birds, butterflies, and other insects.

- Many fragrances emerge at dusk; seed catalogs offer plants that open at night and give off intoxicating scents.

- As seasons change we notice how different the air smells. You can smell newly fallen snow, the buds of spring, and the dryness of leaves on the ground in fall.

Some of the most fragrant plant varieties for your garden:

Perennials

Evening primrose, iris, peony, violet, hemerocalis daylily, hyacinth, phlox, rose, honeysuckle, lily of the valley, tuberose, jonquil, daffodil, jasmine, fall blooming clematis, some hostas, mock orange, lilac, spicebush viburnum, monarda, and scented geranium.

Annuals

Datura, four-o-clock, sweet William, dianthus, sweet pea, sweet alyssum, heliotrope, and blue petunia.

Herbs

Hyssop, lavender, rue, mint, thyme, rosemary, lemon verbena, and lemon balm.

"Nature's music is the elemental sound, living and vibrating from our DNA, the synapses in our nervous system, in the pulsing of our hearts, and the signals sent out by cells and tissues."

Philip Sutton Chard

TASTE

- Taste is the least researched and understood sense.

- Because of chemical sprays, aerosols, and environmental pollution, we have become anesthetized to many tastes.

- Food has long been associated with stimulating sexual interest and some foods, such as chocolate, stimulate endorphins and seratonin.

- Many flowers from the garden are used in cooking, such as roses for flavoring; nasturtiums and pansies in salads; candied violets on cakes; floating flowers in a punch bowl; hibiscus, jasmine, and chamomile in tea; and dried tiger lily buds to flavor Chinese hot and sour soup.

- Herbs have been used to stimulate the taste buds for thousands of years.

SIGHT

- The eyes contain 70 percent of the body's receptors and feed information instantaneously to the brain.

- The setting and rising sun and the change of seasons cast glows on the garden creating fairy realms of color that captivate.

- A winter scene is like a desert with its own form of beauty. Feed your senses even in the winter as you look at your garden with snow on it. Berries that are left on some plants and bushes and the brown stems and fronds of grasses and plants are dressed in their winter garments. The soul takes this in and uses it as fuel to keep it warm until spring.

I have some flowers in my garden just for their form or color and not for their scent. Some of my lilies start out in the morning just after opening up with a yellowish-green tint. As the day goes on they develop a pinkish magenta hue at the edge of their petals that delights the eyes.

78

Intuition

I watch the wildlife coming to drink, play and gather food and materials to build their nests. This activity gives me much pleasure and feeds a place deep inside of me that cries out for nourishment. I also enjoy seeing the rituals of mourning dove pairs feeding on the ground. They mate for life, as do some species of birds, ducks, and geese.

INTUITION: THE SIXTH SENSE

Your intuition is linked to your feelings and is a direct result of a felt sense. "It alerts us that something is wrong when nothing appears to be wrong and that everything is right when nothing appears to be right at all," says Jeff Cox. Many times you have acted on your hunches, sensed you were being watched or experienced déjà vu.

You have a direct perception of the truth when you access your intuition. Using this sixth sense can help you understand your garden on deeper levels and evaluate your experiences there with greater accuracy. The garden promotes flashes of intuition because it fosters emotion, inspiration, and passion. When your environment is harmonious, you can tap into your intuition and locate anything that is out of balance and set it right.

As you engage your intuition it is also important to observe your thoughts. Do thoughts come to you quickly or slowly? Any changes in thought patterns are a barometer of how the space is impacting your energy field.

Observational skills become as important as sensing when working with your intuition. Becoming aware of what your body is telling you and acting on these feelings is also important for developing your intuition. You will find that an open mind, patience, and practice will serve you well for improving this ability.

Intuition is fed by messages coming from the immaterial and unseen world.

79

Feng Shui

Being able to sense energy on your property, feel it in your body and then respond to these feelings are essential to Feng Shui. Feng Shui is the relationship between the energy in your space and the energy in your body. You can then discern what is happening in your garden and your house and relate that to what is happening in your life. (See the chapters on Feng Shui.)

THE SIGNS TO LOOK FOR

Your body always tells the truth, so look for changes in breathing, energy levels, physical discomfort, or emotions and feelings. Quiet yourself and begin to communicate with and observe your emotions and feelings. Intuition may become blocked when you are distracted or chaotic events or experiences come into your life. Low energy levels, inability to focus clearly or a high level of disorganization or clutter near you or in your life also decrease your intuitional skills.

What are your immediate impressions when entering a space?

- Does your breathing expand or become restricted, slow down or speed up?

- Is your breath lively and easy?

- Do you feel restless, tired, calm, uneasy, or off-balance?

- Do you feel grounded, centered, and stable or spacey and disconnected?

- Do your feel relaxed, open, safe, secure, and happy?

- Are there any changes in body temperature or taste?

- Take careful notice of your energy level.

- Pay attention to body reactions of aches, dizziness, and nausea, discomfort of any kind, a tightening in the stomach, or suddenly feeling ill at ease.

On your property:

- Notice if the air smells musty, dry, damp, or dank or has any unusual odors.

- Observe the changing aspects and amounts of light, shade, and darkness.

- Is the vegetation lush and full or dry and sparse?

- Listen for the sounds of insects, birds, and animals.

- Negative energy brings a sense of discomfort or pressure or perhaps something that you can't identify. Assess this too.

The soul loves the rich fabric of the garden because all the senses mingle and are satisfied there. If we did not protect all the beings that saturate the garden and the natural world with their song, fragrances, visions, impressions, and flavors, we would lose a great deal.

The garden is an ongoing creation.

"Like rain upon the fields, the soul seeps deeply into all things animate and inanimate. It is these drops of soul, embedded within the objects of the world, that give life and vitality to existence."

Eliezer Shore

81

ANGELS AND FAIRIES

"Soul is where the fires of our passions burn. It is where our love is most alive. The soul longs for this deeper love, for a connection between form and formlessness, for a continuum between the earth and the divine."

—Benjamin Shield, Ph.D.

In Greek angel means "messenger of god." Angels are also described as ministering spirits, guides, and guardians or protectors. They have decided not to reveal themselves, although many people claim to have seen one. Acknowledging angels and spirits allows us to feel and experience life on a more spiritual level. Surveys have shown that more than 85 percent of our population believe in them. Faith in beings that can come to our aid and who bring an element of the divine into a very materialistic world offers hope and optimism. They are part of every religious and spiritual tradition on Earth. Our ancestors called them by their names. To have angels in our lives requires having our imagination regularly enriched by stories and artistic images that teach us how angels look, how to call them, and to know when they are present. In our culture today, we have movies, TV programs, songs, and books that have shown how the presence of angels can change lives.

Angels are beings hidden behind the work of creation; it is the nature of divinity to remain invisible in human affairs. They help us connect with the divine, act as intermediaries, and are co-workers or

"The angelic forces are concerned more about your willingness to actively participate with them in maintaining the perfection of creation than in the mere 'mouthing' of words which may make them appear holy."

P.M.H. Atwater
Mighty Natural

HOW TO MAKE CONTACT WITH AN ANGEL
It is important to be in a prayerful, meditative,
or spiritual state of mind to begin contact with the angelic world.

1. Sit in a tranquil and quiet spot and totally relax the body and mind.

2. Have a clear image in your mind which angel you want to reach and why.

3. The goal is to gain clear, insightful direction and guidance after you pose your question.

4. Relax even more, deepen your meditative state, and just wait. Listen without expectations.

5. Initial contact may only be a subtle feeling that a powerful presence has entered your awareness.

6. For most people more than one try is necessary.

7. Don't expect to hear them; they communicate by helping you see and feel the answer to your question. You may become aware that there is a presence that is communicating with you. Note: Communication is used only to describe the fact that a question has been asked and that an answer is expected in whatever form the being you asked chooses.

8. Remember to thank your angel.

co-creators with us to bring about a more spiritual expression of life on this planet. There is an over-lighting angel for each and every garden. Each one embodies a particular quality, radiance, vibration, or life form.

When you talk to them, you might ask what to do about your deer or ant problem, or where to place a plant, or why a particular tree is ailing. There is ample evidence in literature to assure you that you can contact Nature through the angels, ask questions, and expect to receive answers. Allow the answers to flow into your mind as ideas and images. It is good to thank them for their help. You will feel what is right for you.

Five things are needed to contact these beings:

1. An open heart and mind.
2. Purity of intention.
3. Willingness to regard Nature with respect.
4. A sense of cooperation in working with them.
5. Belief in their willingness to work with us.

FAIRIES

Deva means angel in Sanskrit. Some sources use angel and deva interchangeably.

When we were children we talked and played with fairies and felt close to this realm. Why did we stop? Why are the cultures of Scotland, Ireland, Wales, and England more aware of and involved with the "wee folk"? If one were to bring up the subject of the "wee folk" over there, you would get a circle of people around you interested and sharing stories, all believing in them. In the United States we sense an embarrassment when they are mentioned, as though it is childish. Is it our puritan heritage?

The fairy realm is a timeless place of regeneration, beauty, allure, and wisdom. The counterparts of all animals, birds, fish, insects, trees, plants, and all living forms that exist on this planet are found there. It exists simultaneously with ours and is an archetype of the natural world.

The fairy tradition traces its roots to many religions. The materialistic nature of our culture cannot accept these beings because we are unable to prove the existence of this otherworldly place. The fairy realm can help us regenerate our love of Nature and transform our lives for the better; at the same time it is felt that one must obey some simple rules of engagement when relating to them. Use the steps mentioned above on how to contact angels.

"Everything God created is potentially holy—everything has God's fingerprints on it."

Rabbi Harold Kushner

Gardens connect you to the soulful longings of your heart.

Fairies can be mischievous but are not evil. They can take any form that you expect, dream of, or desire. If you suddenly are aware of a lovely fragrance, a soft rustling in the tree you sit under, a sudden whirling circle of leaves, a gentle breeze, a shimmering light, or a shadow that can't be explained, it is possible that a fairy or even an angel has revealed itself to you.

Music attracts them, as do personal tokens, cookies, small amounts of hair, shells, rocks, or crystals. These may be put out in shady spots, near mushrooms, on rocks or mossy patches, near waterfalls, or anywhere in the garden that catches your fancy and that you have set aside for the purpose of attracting fairies. Also leave notes for them in the holes in trees. Believing in these beings will encourage them to appear. They love honeysuckle, thyme, Johnny jump-up, foxglove, primrose, bluebell, cowslip, and Harry Lauder's walking stick as well as running water. Creating a spirit path or walkway is another way to bring the nature spirits to your garden. Leave a wild place for them somewhere in your garden; they feel more comfortable in the wild.

At Findhorn, Scotland, in the early 1960s several people banded together on a remote, barren, and windswept stretch of land to begin an experiment in living and working with what they called nature spirits or intelligences, devas or overlighting angels, employing these terms interchangeably. They learned to communicate, pray, and meditate with the angels of the land and plants. Within a few years, they

AUTHORS NOTE:

Some confusion about who to call for what.

In the articles and books I have read it was stated that fairies take care of individual plants, trees, and other life forms, such as cowslips, elms, or parsley. Angels are concerned with overseeing all of Nature and responding to calls for help from humans. I did discover that some sources use angels and fairies interchangeably as beings to contact whenever specific help is required. When you desire an answer about an individual plant or life form you may ask the angel or fairy of that particular thing. Please use your own guidance in these matters.

The soul connects us to the eternal and divine source of life, love, and the essential good in the world that weaves all things into one fabric of life.

were growing cabbages the size of basketballs and have since established a school and residence for those interested in furthering their knowledge about working with these beings. They became aware of plant consciousness—indeed the consciousness of all living beings—conversed with and used their advice to change a barren spot into a paradise.

At Perelandra in Warrenton, Virginia, Michaelle Small Wright has perfected some other methods of communicating with nature spirits that include kinesiology. Her particular form is holding the pinky to the thumb, asking a question, and trying with the fingers of the other hand to break the finger bond. When the answer is "yes" the bond will hold, when the answer is "no," the bond will break. She raises many vegetables, herbs, and flowers that are then used to make essences for healing, teaches classes, and has published books about her work.

The success of Findhorn and Perelandra and your ability to contact angels and fairies lies in establishing a partnership with Nature. It requires communicating on many levels with the spiritual essence of Nature, beginning to think like a plant, talking with them, and believing that the unseen world that created all the forms of Nature also created us.

The ancient cultures of Egypt, Babylonia, and Greece, as well as Hawaiians, Native Americans, and others, believe air, wind, and water have personalities and intelligence, and are alive, real, and approachable. Indigenous cultures all over our planet have accepted these powerful energetic forces and endeavored to work with them, not ignore or deride them. It is important to realize that a belief in spiritual beings and that Nature has intelligence does not necessarily negate using science to better the world.

Angels and fairies are time-honored beings that the soul loves to communicate with and be with in the garden. Since most of us believe in these beings anyway, why not explore a relationship with them in your garden? They personify the highest form of soul qualities and are aligned with beauty, divinity, and the eternal.

Garden Gift: CALENDULA SKIN SALVE

For skin irritations and burns including sunburn

1 cup fresh calendula leaves	$\frac{1}{4}$ cup almond oil
$\frac{1}{4}$ cup melted bee's wax	$\frac{1}{4}$ teaspoon borax
$\frac{1}{4}$ cup lanolin	$\frac{1}{2}$ cup boiling water

Directions: Pour water over leaves and let steep until cool. Warm the oil, wax, and lanolin in a double boiler. Dissolve borax in this mixture and then add the leaves. Mix all until the consistency of heavy cream. Cool and then pour into a glass container.

Family Circle, Easy Gardening, spring 2001.

FENG SHUI BASICS

"Like every explanatory system science is partially mistaken. To put it another way, science gives only a partial view. Faced with a crisis that has its roots in science we must open ourselves to other ways of knowing."
　　　　　　　　　　　　　　　　　　　　—John Brookfield

Feng Shui (pronounced fung shway) is a Chinese system of centuries-old wisdom that uses the practical, the aesthetic, and the mystical to forge a pathway to wholeness and a deeper meaning of self. Feng Shui also incorporates the use of colors, shapes, and symbols to enhance and strengthen the positive energies on a site. It is always mindful of how one's environment and the things in it affect the person. Spirituality and purity of intention further connect this system to soul-based purposes.

Using Feng Shui you can find a place where energy flows smoothly and yin (feminine, dark, and slow) and yang (masculine, light, and fast) are in balance. Its main purpose is to establish harmony between the dweller and the dwelling and to create nurturing and supportive environments.

CHI

The concept of Chi is central to the study and use of Feng Shui in one's environment. Chi is the universal invisible life force underlying all life's processes. Everything is connected by this energy, which even

"Gardens are boundary lines of a transition to new realms of experience, mystery, and spiritual awakening."

Anthony Lawlor

appears in inanimate and man-made objects. It moves in curves and spirals like water, as it ebbs and flows everywhere. Since we are surrounded by Chi and it exists in all things, our gardens are also affected. Once we learn to sense positive and negative energies on our property, we can correct imbalances and have more tools to bring soul to our spaces. We sense something called place energy. The chapter on Senses has more information related to this topic.

Reducing or doing away with negative influences encourages positive forces to remain on a property and creates opportunity, good health, and luck for the inhabitants. Positive Chi feels smooth, flowing, uplifting, and powerful. Negative Chi feels rushed, blocked, stagnant, piercing, or draining.

10 BASIC TOOLS TO ENCOURAGE BENEFICIAL CHI IN YOUR GARDEN

1. Colors
2. Sound makers
3. Mirrors
4. Crystals
5. Lighting
6. Art
7. Living things
8. Water features
9. Moving objects
10. Natural objects (rocks and shells, for example)

SOUL IN A FENG SHUI GARDEN

Feng Shui brings soul to a garden in subtle yet profound ways by using gently rounded features that mirror the way Chi flows in Nature, a balance of plant forms, colors, and textures, and seeding every project with a sense of the sacredness of Nature.

Feng Shui honors the earth as a living organism that is nourished by forces such as the wind, sun, moon, rain, and seasons. It is also about how the environments in which we live, work, and play have an effect on our physical, mental, emotional, and spiritual well-being. Using the Bagua, the five Elements, concepts of the flow and nature of Chi, and symbols help us to understand that our homes and gardens are direct extensions of ourselves. They are mirrors reflecting who we are, where we are going, and who we wish to become.

The Chinese believe the human must create harmony in his/her life so it can flow to the rest of creation. Inner harmony leads to order and harmony in the world and has a way of filtering down from the individual to family and then into society, country, and so on. They believe there is a correct way to live and that how you live ultimately affects your health. Feng Shui uses the Bagua to help generate the best possible environment for well being, glowing health, comfort, safety, and happiness.

THE BAGUA

The Bagua is an ancient Chinese symbol and tool for explaining life issues and desires. It is composed of nine sections and is an octagon, which is a sacred symbol to the Chinese and many of the cultures in the Orient. It describes aspects of life and then positions these in a template covering a home or garden. Looking into the various areas you find life descriptions, issues, colors, and elements. Deciding what to work on in your life, going into the garden and finding the matching areas of the Bagua, are the first step. The second step is creating enhancements using decorative objects or choosing plants in corresponding colors and shapes of the particular sector you have chosen to work with. It is hoped this will bring about the change you desire. In Feng Shui we enhance the outer environment to stimulate the inner environment.

To begin your work, determine into which doorway of the Bagua

91

The Bagua

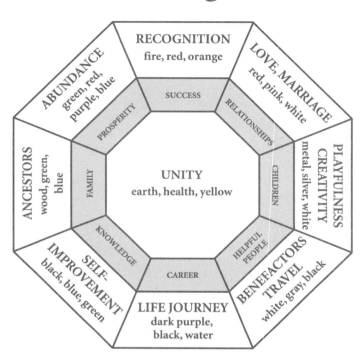

- **RECOGNITION** — fire, red, orange
- **SUCCESS**
- **ABUNDANCE** — green, red, purple, blue
- **PROSPERITY**
- **LOVE, MARRIAGE** — red, pink, white
- **RELATIONSHIPS**
- **ANCESTORS** — wood, green, blue
- **FAMILY**
- **UNITY** — earth, health, yellow
- **PLAYFULNESS CREATIVITY** — metal, silver, white
- **CHILDREN**
- **KNOWLEDGE**
- **HELPFUL PEOPLE**
- **SELF-IMPROVEMENT** — black, blue, green
- **CAREER**
- **BENEFACTORS TRAVEL** — white, gray, black
- **LIFE JOURNEY** — dark purple, black, water

The size of the square Bagua can expand or decrease to fit the dimensions of your garden

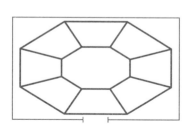

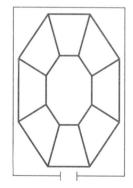

you enter your garden, garden room, or designated space. This marks where the rest of the sectors fall. Create a Bagua wherever you want in your garden. Have fun with this concept.

WAYS TO USE ENHANCEMENTS IN YOUR GARDEN BAGUA
(You may look back under Chi to view the 10 enhancements.)

CAREER/LIFE JOURNEY
Place a water feature or birdbath near your front door or in this section of your garden. This will bring new life to your career and prosperity to your life.

HELPFUL PEOPLE/ BENEFACTORS/TRAVEL
Birds are lively messengers making birdhouses and feeders appropriate here as well as statues of Buddha or other positive figures or symbols.

CREATIVITY/CHILDREN
A good area to place your potting bench along with pots to start seedlings.

RELATIONSHIPS/MARRIAGE
Friendships are made and nurtured in the garden; placing a table and chairs for intimate

tête-à-têtes goes well in this area. A bench outside the master bedroom is good for fostering romance, as are using pairs of objects.

SUCCESS/RECOGNITION

A path leading to a focal point will bring clarity and illumination.

Prosperity Wealth Fortunate Blessings Abundance	Fame Reputation Success Recognition	Relationships Love Marriage Friends
Family Elders Community Ancestors	Center Unity Health	Creativity Children Projects Future
Knowledge Skills Wisdom Self-Improvement	Career Path in Life Journey	Helpful People Benefactors Travel

Decide which door is the entrance to your garden.

Use the square Bagua to place over your garden. We have made the octagon fit into this shape because it is easier to work with since most spaces are square and not an eight-sided figure.

WEALTH/ABUNDANCE/PROSPERITY

Your compost heap is wonderful here as it represents black gold. Nine fish (eight gold and one black) in a pond also bring prosperity. Try planting gold colored flowers in this area.

FAMILY/ANCESTORS

This is a good area to create a remembrance garden where plants from family gardens are placed to bring back memories of good times and the people you spent them with.

KNOWLEDGE/SELF-IMPROVEMENT/SKILLS

A meditation spot is ideal here.

HEALTH/UNITY

This is usually the center of the Bagua and represents infinite possibilities, opportunities, and unity. Be mindful of its importance in your life and think about installing a medicine wheel, labyrinth, or guardian rock in the center, or a meditation or spirit path leading to the center. Leaving it open and free is also a choice. Health issues may also be addressed in the FAMILY sector by planting from the list in the chapter called Symbols.

"One of the hallmarks of a healthy mind is gratitude and thankfulness for Nature's Gifts."

Thomas Moore

MORE BAGUA CONNECTIONS

Following are additional concepts to consider when using a sector of the Bagua that applies to what you wish to work on in your life.

Career: Life purpose, goals, beginnings, self-appreciation, communication, emotions, discovery, willpower, and drive.

Helpful People: Nurturing and support from others, visible and invisible means of support, and networking.

Children: Joy, projects, nourishing the inner child, being grounded, and fun.

Relationships: Intimacy, sharing of self, commitment, feminine energy, coziness and softness.

Success: Personal accomplishments, honoring self, self-expression, illumination, clarity, vision, finding your way, and visibility.

Wealth: Empowerment, resources, needs met, freedom from limits, orderliness, opportunities and invitations.

Family: Roots, past history, tradition, your place among others, and holding on to stuff from the past.

Knowledge: Dreams, self-discovery, spiritual growth, inner knowledge, intuition, new relationships, and fertility.

Health: Nourishment of body and soul, coming full circle, feeding one's visions, and encouraging a holistic view of life.

THE ELEMENTS

Feng Shui offers us new tools to enrich our environmental experiences. In a Feng Shui garden it is important to use the traditional Chinese concept of the five elements: Fire, Earth, Water, Metal, and Wood. A greater degree of balance and harmony will be achieved in any space if they are used in the form of plant shapes, color, and objects that reflect the element. Look in each sector of the Bagua for its element. Adding little touches to your garden using this concept will give your space a completed, stable, nurturing, and vital feeling. Each element has a unique geometric shape assigned to it. "We receive messages from the things around us, including geometric shapes and patterns. It is programmed into our cellular structure and memory that we yearn for the natural geometric shapes that underlie all Universal order," says Denise Linn, in *Feng Shui for the Soul.*

> *The well-being, health, and prosperity of the occupants of the property are as related to the house as they are to the garden.*

FIRE: An active, vibrating, pulsating energy that is associated with the chemical transformation that occurs when cooking. You can use red paving bricks in angular patterns. All lighting, candles, fireplaces, barbeques, pets, wildlife, all art that depicts animals and people, pyramidal and triangular shapes, and the color red symbolize fire. Use flowers in warm colors.

EARTH: The energy of earth is gathering and downward, and it symbolizes the settling energy of the late afternoon, the ripening of crops, the gathering of the harvest, and the process of maturation. Earth represents the center. Brick, tile, adobe, square and rectangular shapes, long, flat surfaces, and the earth colors of yellow, gold, orange, rust, and brown are examples of earth energy.

WATER: The essence of water is persistent, fearless and always finds a way around obstacles. It is evoked by the quiet darkness of night and the stillness of winter. The continuous movement of water represents new ideas coming forth. Irregular, asymmetrical, and free form shapes are symbolized by water, as are reflective surfaces, glass, crystal, mirrors, and the colors of black, deep blue, and dark purple.

METAL: Metal represents energy that moves inward and is solidifying. It symbolizes the completion of the harvest and the end of

C

> *"Care of the soul leads to paradoxical mysteries giving poetry to everyday life."*

Thomas Moore

summer. Metal garden art, wind chimes, fencing, and furniture, rocks, sand, arched forms, ovals, circles, and the color of any metal, as well as white, gray, and pastels can be used.

WOOD: The element of wood is represented by the upward and outward moving energy of a tree and associated with growth, vitality, and renewal. Trees, all plants, wooden furniture, decks, fencing, lattice, striped and floral fabrics, columns, long, thin, or tall shapes and the colors blue and green also symbolize wood.

You have begun your journey using Feng Shui. It is an exciting and rewarding complement to other modalities you might use in your garden. Many of its concepts favor a common sense approach to making changes and enhancements that can bring new vitality, happiness, and prosperity to your life.

"See the world as yourself. Have faith in the way things are. Love the world as yourself, then you can care for all things."

Tao Te Ching

🍃 🍃 🍃 *Garden Gift:* 🍃 🍃 🍃
PRESSED FLOWER NOTE CARDS

You will need good quality paper (can be obtained from hobby stores or artist supply stores) and clear glue. You can also use purchased note cards. Gently separate flower petals and leaves from plants you have in your garden. Place these between two pieces of paper towel and put inside a heavy book for at least two weeks. Arrange these on a piece of paper until you achieve a pleasant design. Apply glue sparingly to your chosen papers or cards and then place your petals and leaves from your preformed designs into the glue. Dry well.

Family Circle, Easy Gardening, spring, 2001

SYMBOLS

"To ignore the vitality hidden within the symbols that surround us is to cut ourselves off from the healing and renewing powers of the soul."
　　　　　　　　　　　　　　　—Anthony Lawlor

Symbols are very important to Feng Shui and Chinese culture. We are surrounded by symbols; the flag, McDonald's arches, and car logos, come to mind. Carl Jung describes them as pictures, art, or material objects and icons that possess special connotations in addition to their conventional and obvious meanings. Many of these archetypes are strikingly similar in cultures around the world and even appear in our dreams.

The image of the Sun as used for decoration in so many gardens may symbolize unity (from the circle or wheel shape), life-giving warmth, happiness, joy, optimism, and being full or complete. When you place it in your garden you will feel the effect of what it symbolizes immediately. All symbols enrich your surroundings with subtle meanings, have the power of suggestion, and stimulate your subconscious mind reaching you more powerfully and subtly than rational linear thinking does.

Unpleasant views can also affect your life. When your surroundings are not nourishing, your interaction with the world is impaired. Living in a beautiful environment is much more supportive for everything you wish to accomplish than unattractive, broken, uncared for, and messy scenes which symbolize a level of

"Nature's melodies—created by the waters, wind, animals, sky, plants—sing in a cadence that we already know deep in our souls, but have often forgotten."

Philip Sutton Chard

disorganization and blocked energy. You will feel more powerful and confident if you leave a clean, esthetically pleasing, and balanced environment, than a messy, unattractive, and chaotic one. Looking forward to coming home to a harmonious setting can make your day a joy and represent to you that your life is bountiful and rewarding.

Boundaries such as fences, walls, hedges, or entrances and gates are symbols for transitions. They also symbolize and provide safety and protection. Plant hedges or bushes or build a fence near the house to counteract steep grades that fall away from the house. This can represent money leaving. Prepare a butterfly garden in a sunny sheltered area of your garden. Since butterflies are an ancient symbol for the soul their presence in your garden will nourish your inner self. Bring the symbols that you want to have in your life by planting trees, flowers, bushes, and herbs that have specially acknowledged meanings such as:

Pine: resilience, integrity, and dignity.

Yew: protection.

Plum: friendship.

Pear: health.

Peach: brotherliness.

Apple: safety and well being.

Orange: good fortune, generosity, happiness, and wards off bad luck.

Cherry: youthfulness and spring.

Bamboo: fidelity, wisdom, humility, peace, balance.

Willow: gracefulness.

Magnolia: beauty and gentleness.

Oak: long life, protection, and steadiness.

Olive: peace.

Pomegranate: fertility.

Apricot: beauty and charm.

Mum: long life.

Peony: prosperity, love, and marriage.

Jasmine: friendship and affection.

Lotus: purity, and inspires peace and contentment.

Hollyhock: fertility.

Daffodil: open-minded and generous.

Geranium: prosperity.

Rose: love and romance, hope, and youthful beauty.

Water lily: purity and truth.

Jade, money plant and silver crown: prosperity

Using positive symbols can increase the amount of harmony in your garden and will encourage it to flow into your life. Symbols are experienced as subliminal messages in the garden that bring to conscious awareness desires, goals, and feelings we wish to materialize in our lives.

USING APPROPRIATE SYMBOLS INCREASES HARMONY IN THE GARDEN

Dreams are the symbolic language of the subconscious mind.

1. Take a look at your garden to see if there are any images or objects that speak of disharmony, neglect or being crowded, overlooked, unattended, out of control, dead, or broken.

2. Determine if any of these situations mirror events in your life at the moment. Your environment should express the positive and best in life.

3. List the healing qualities you desire in your life and find symbols that remind you of these.

4. Check your body reactions for clues about how these make you feel. (See Intuition, in the chapter called Senses for more ideas.)

5. Become acquainted with your own personal symbology. What colors, shapes, sounds, textures, and sights mean security, peace, and happiness to you?

6. Look at the Bagua and decide what areas of life you desire to work on. Then place colors, shapes, and objects that symbolize these in those sectors of your garden that correspond to that sector of the Bagua.

Perhaps the reason we like birdhouses so much is that they are symbols of security, comfort, nurturing, and coziness. You will be

"Some people tell me I invented the sounds they call soul—but I can't take credit. Soul is just the way black folks sing when they leave themselves alone."

Ray Charles

Bird house

Bird house

affected by these messages every day if you incorporate them into your garden. When you see a messy desk, a pile of clothes or newspapers, or flowers and plants that need trimming and deadheading, they announce that your world needs attention and care. If you allow things to get out of control, your energy field gets clogged and choked. You may begin to feel tired, your thinking becomes fuzzy, and decisions are harder to make.

Symbols are around us and with us all the time as we dream and experience our environment on many levels. The subconscious mind always knows what a symbol means because it translates information received from everyday objects, events, memories, and experiences into feelings in your present life and retains every bit of information that you are exposed to. These can exert a powerful force to aid healing, growth, harmony, and balance.

🍃 🍃 🍃 *Garden Gift:* 🍃 🍃 🍃
AROMATIC FIRE STARTER

Gather some twigs from your yard about 12 to 18 inches in length. This provides the kindling for your fire. Cut handfuls of long sprigs of rosemary, heather, lavender, thyme, juniper, sage, and any other aromatic herbs growing in your garden. Wrap the twigs together with a strip of newspaper you have folded into a 4-inch band. Close tightly with masking tape. Place herbs on top and tie with raffia. As the flames heat up in your fireplace a wonderful aroma wafts into your home.

Family Circle, Easy Gardening, spring 2001.

FENG SHUI GARDENS AND LANDSCAPES

"Bad gardens copy, good gardens create, great gardens transform. What all great gardens have in common are their ability to pull the sensitive viewer out of him or herself and into the garden, so completely that the separate self-sense disappears entirely, and at least for a brief moment one is ushered into a non-dual and timeless awareness. A great garden, in other words, is mystical no matter what its actual content."

—Ken Wilbur

A Feng Shui garden is brimming with soul because it fosters positive Chi, strives for balance and harmony, and acknowledges the sacred in all of Nature. Feng Shui looks to Nature for its lessons and wisdom. Over many centuries of use, masters have recorded what works best for humans and their interactions with their environments.

IN A FENG SHUI GARDEN:

Using Feng Shui we aim to achieve gardens and landscapes that echo the natural and organic shape of the land and the surrounding environment. Topiary is not used because it does not follow the natural shape of the plant, bush, or tree. Formal, angular, or boxed-in types of gardens are not encouraged, since plants do not grow this way in the wild nor do these shapes allow Chi to circulate freely.

"Take time each week to study a flower, catch a snowflake, or contemplate a sunset."

Country Living
January, 2001

101

Contrasts—hard and soft, rough and smooth, long and short views—and mixed shapes, sizes, colors, and textures of plants are used because this follows Nature's pattern.

Use the cooler colors of blue and green to balance the hot colors of red and orange. Silver or metal colored plants stimulate prosperity, such as gold spirea or dusty miller. Balance the soft, yin qualities of ferns with the harder upward thrusting yang qualities of hibiscus.

Overcrowding is not desirable because the eye cannot distinguish individual plants. Allow each plant to speak for itself. Every plant is treated with the utmost respect. It is best not to use thorny plants such as roses, cacti, yucca, or agaves near the front door. There is a potential for being snagged. Some roses do not have thorns, such as Zephirine Drouhin and are acceptable at the front door. If you live in the southwest, round cacti are often used by the front door, as Native Americans do. I believe this is acceptable because one has to accommodate to their surroundings.

We can enhance missing areas by filling in incompleted geometric forms that occur in home shape, lot configuration, and garden layout with gardens and landscaping. It is advantageous to create harmony in the landscape by using flat areas to balance mountains, hills, or tall trees.

Entrances are the major inlets of Chi in a space.

ENTRANCES

Entering a new space may be accompanied with some fear of the unknown, unseen, and mysterious. In ancient cultures, gates and entrances were given respect and symbolized aspects of the sacred. People were taught to bow down and kiss the earth of a garden or farm, showing reverence and gratitude for the gifts of life provided so sumptuously within.

- Entrances and gateways possess an air of mystery; we wonder what is beyond. Our curiosity is aroused, and we enter with a heightened level of expectation.

- Entrances and gateways signal a passage to new realms of experience, pleasure, and spiritual awakening, forming a bridge from the outer world to our inner world.

• When passing through the gate or entrance to a garden we enter a special world where time seems to stand still or at least appears to move more slowly. How many times have you set out to garden for a few hours only to discover that an entire day has passed without any awareness of time?

• When we pass through the entrance to a garden, we enter a special world that is safe, nourishing, comforting, aesthetic, and carefree.

• Our nerve centers are activated and our brains begin to produce hormones and chemicals of pleasure.

• Gates, arbors, trellises, archways, and doorways of all kinds to gardens are beguiling and give an air of permanence while welcoming and inviting in visitors. Arbors bring back romantic images of one's marriage or other positive events.

• As you look in all directions it is desirable to encounter plants, flowers, a winding path, and a distinctly marked and well-maintained front door. Your subconscious mind takes this in and interprets it as safety, nourishment, comfort, and a welcoming sign. Even before you come into the house, you have a sense of hospitable cheer and potential for pleasant experiences.

Some Tips For Garden Entrances:
Creating a path that meanders and winds encourages people to slow down and leave the work world behind as they approach the entrance. Different types and sizes of pebbles and flat stones used

What the Soul Needs:
• *A connection to the land*
• *Feeling safe*
• *Harmony with the greater cycles of nature*
• *Sacred space.*

Denise Linn

artistically will add to the textural variety and soften the straight lines of most walkways. Plants of varied heights and textures can be used nearby to give the eye a visual treat. A sitting area near the entrance draws the eye in and helps to create a focal point. Garden art, a birdbath, water feature, or bright annuals will also accomplish this. A comfortable sense of enclosure is created when potted plants, rock arrangements, flower boxes, vines and mood lighting are used near an entrance, gate, arch, or trellis. These often frame the view ahead, providing a sense of mystery and enticement. A sign that welcomes visitors and helps set the tone and purpose of your garden is also a good addition.

Each one of these suggestions may also be used for the front entrance of the house.

PATHS

Experience your garden as though it were a doorway to your spiritual quests in life.

- A path approaching the home, if moving slightly upward, is a sign of good fortune. A path going downward a bit as you leave the house indicates life is easy. A path that narrows to the door acts as a funnel and concentrates too much Chi toward the entrance. Too wide a path can dissipate the Chi over a wide area, weakening it before it enters the house.

- Walkways should be wider toward the driveway or the street and ideally are meandering curves.

- Straight lines are rarely found in Nature, cause a feeling of being rushed, and are like arrows speeding toward their target, which give us an uncomfortable and threatening feeling. Nevertheless, in the garden and in pathways they can be used occasionally to move energy where it is needed. If paths or beds contain too many sharp angles, Chi movement can be disrupted. Rock gardens and plants can slow down the movement of Chi, if that is needed.

- Paths invite you to slow down, observe, focus on, and savor every tantalizing treat along the way—the scent of thyme, roses, violets, or a view of romantic images.

- Create a meditation or spirit path meandering through your garden and fill it with pine needles, gravel, or stepping-stones while using ancient archetypes and symbols along the way that have special meaning to you.

STAIRWAYS

Stairways are also an important consideration and should be safe, not too steep, and free of any sharp angles or turns. It is important to consider having several levels of transition to the porch and front door. This can be accomplished by adding a series of meandering beds to catch one's attention.

FRONT LANDSCAPING

Keep front gardens simple, while representing your personality. A crowded front garden can catch too much of the Chi before it enters the house. The path to your front door is one of the most important elements to consider and must be clearly marked as you enter the property to avoid uncertainty. Be careful that it does not have any overhanging plants that restrict movement. Stand at the front of your home and note how it makes you feel. Your view of life is directly affected by what you see every day. It is desirable to remove unpleasant views, crowded plantings, all dead limbs and plants, overhanging branches, and overgrown areas.

I recommend cleaning up messy front porches. This presents a physical obstacle each time you come into your space and creates psychological impediments to movement in life. Bikes, toys, boxes, trash, and so forth near the front entrance, on the porch, landing or for that matter anywhere on the property, are dangerous and unsightly, and create blocked energy, limit your opportunities, and cause fuzzy thinking, frustration, and confusion.

"There is a great dream across the world that we are a part of. It is not like any ordinary dream in sleep; for we do not dream this dream. Instead this dream dreams us. It dreams us all the time, even while we are awake, and we know that it must be lived on earth through everything we do."

Bushman of the Kalahari

105

It is important to eliminate all multiple doors, gates, and porches from the approach to your home and make sure there is a walkway from your front door to the street. Misaligned doors are like unbalanced wheels in a car. These can symbolize stumbling blocks to your path in life.

Planting hedges and installing berms or low fences at the front of the property can achieve a sense of protection and security. This is an especially important consideration if the home faces a busy street. The noise and fast-moving Chi can disturb the residents and cause opportunity and luck to pass by.

It is desirable to "create an embrace of landscaping around all sides of the house which can also help to balance extreme architectural features such as sharp corners and hard surfaces," says Terah Kathryn Collins in *Contemporary Earth Design*.

TREES

- Trees are important because they provide us with oxygen and vitality, and impact our psyche with images of growth, renewal, and hope.

- It is best not to have too much shade or sun in the front. Strive for a balance.

- Trees also soften the foundation and sharp corners or edges of the house. They frame the house and tie it more closely to the ground, as well as accenting and calling attention to the entrance.

- Trees in the back act as a mountain protecting the house and the property. This puts the house in the desired "arm chair" position. One can also create a mountain by using a bank of bushes or a raised berm of earth.

- Avoid planting trees that block the front entrance because this hinders your approach to the house and can create obstacles in life. If a beautiful old tree is blocking your front door it is not necessary to cut it down. A remedy for this consists of writing "raise head, see happiness" on a red circle on a new pad with a new pen and posting this at eye level on the tree, says David Daniel Kennedy. Do not use a nail to fix it to the tree. Do the three secrets reinforcement technique on page 110 or say a prayer after this to ensure good results.

- If any tree roots grow near the foundation or cut off the plumbing, the tree must come out, or its roots pruned.

- Chi will be trapped around one lonely tree, as it becomes a dominant feature taking away from other beneficial elements.

- Consider using various forms of trees: some are pyramidal or vase-like; some are weeping and form a curve or oval shape. You can use the shapes of trees to augment your use of the elements on your property.

- Weeping tree forms are best used in the back yard as they could foster a sad feeling when you enter. When planting a tree take note of its symbolic meanings and plant those that express a feeling you wish to welcome into your life. See the chapter called Symbols for some suggestions.

Consider the shape of the land and incorporate its natural contours in your landscape, garden, and house plan. This will cut down on the amount of soil and green cover destruction. Considerations of ecology, conservation, and preservation of the natural beauty and characteristics of the land are of major concern in Feng Shui because its Chi is revealed this way.

The Earth inspires our creativity with its myriad forms, colors, shapes, sizes, textures, smells and visions of enchantment that feed our sense of beauty.

🍃🍃🍃 *Garden Gift:* 🍃🍃🍃
WINTER WARM HERBAL BATH

2 cups dried rosemary
2 cups dried or fresh lavender
1 cup thyme

1 cup bay leaves
1 cup calendula flower petals
1 cup rose geranium leaves

Store this mixture in a cool, dry place in a sealed container. Blend the herbs well before using.

Directions: Cover ½ cup of the blend with 1 cup boiling water and steep for 20 to 25 minutes. Drain liquid into the bath water. After a long soak use the washcloth as an herbal scrub. The remaining mixture is enough for about 15 more soothing baths.

Family Circle, Easy Gardening, spring 2001.

WATER IN THE GARDEN

The ancient Chinese believed that the flow of water mirrored Chi currents that surround the Earth. Lillian Too says that when the main door faces water or there is a view of a body of water, the Chi from the water creates the potential for prosperity to flow into the surrounding areas. Water is an important element in a Feng Shui garden because it brings vitality, attracts beneficial Chi, and adds an element of calmness to the environment. Because water represents the feminine it can be used as a balance for the masculine elements in a landscape, such as rocks, cliffs, and wide vistas.

Swimming pools or ponds should be oval, round, or kidney shaped. It is important their size not overwhelm the house. Water features, being yin, balance the overall yang qualities of the surrounding foliage. It is desirable that water be kept clean and moving slowly toward the house in a curving manner. This encourages prosperity and abundance to enter the lives of the occupants. A stream or river, flowing behind a house represents missed opportunities.

Place a water feature to the left of the front door, as you look out from

inside your house. Avoid a water feature in the fame or marriage area. Waterfalls and sprays create negative ions in the form of mist, which is excellent for your health and psyche. Fish are symbols of prosperity. Try to use eight orange and one black in your water feature as mentioned before.

RECOMMENDED FEATURES

Thresholds, depth views, and focal points are also important considerations as they enhance aesthetic qualities and draw Chi in to linger awhile. Garden rooms that beckon and increase the element of mystery are good to work into your overall garden plan.

Materials and shapes used in the building of the house can be repeated in the garden, as they provide connection, relationship, and easy transitions from house to garden.

Add rounded flowerbeds around a rectangular patio to soften its hard edges. Let your landscaping shapes take their cues from cloud shapes that you see in the sky. The natural world rarely shows a sharp edge. You will find the most pleasing gardens are ones with lots of curves. Curved lines encourage Chi to flow more freely and without obstruction, allowing more to get to you and to enliven your environment. Create meandering lines everywhere to follow the curves in the land and mirror the way Chi flows.

Add a bench to your garden for moments of quiet contemplation. Plant for tranquility using pastels, creams, or soft greens that will help to create an oasis of calm in a busy world.

Plants of various sizes, shapes, and textures grow together in Nature. A balance between sun and shade is important since sun represents yang and shade represents yin. Consider thinning out your trees if there is too much shade to let more sunlight through, or plant a stand of bamboo to create a quick solution for too much sun.

"Simple living integrates both inner and outer aspects of life into an organized and purposeful whole."

Duane Elgin

109

EDGES

Edges are important in a Feng Shui garden because they represent the interface between two systems. Chi tends to accumulate there. Some edges receive more light and nutrients than others and are therefore more productive. In Nature, we find edges at shorelines, forests, and places where mountains meet valleys and plains. These sites have particularly high levels of Chi and beauty that feed our soul and inspire us. In the garden, edges appear at the rim of a water feature and at the borders of flowers beds and landscaping.

LOOKING AT YOUR GARDEN FROM INSIDE YOUR HOUSE

As you go about your daily activities of eating, working, cleaning, cooking and so forth, it is nice to have pleasant and inspiring scenes to view. I suggest creating spaces in your garden that can be seen from the house.

- Place bird feeders and birdhouses where you can see the activity taking place.

- Position a water feature so you can see the wildlife that arrives and hear the calming sound of trickling water.

- Establish theme gardens of romance, remembrance, relaxation, meditation, healing, or prosperity that bring their message to you whenever you glance out your doors or windows.

A SHORT GUIDE FOR CREATING
A FENG SHUI GARDEN

Elements: Be sure to include the five elements by employing their shapes or color in decorative objects or plants. A square glass table (water and earth) with wrought iron base (metal) integrates earth, water, and metal. A round water feature with rocks, surrounded by plants represents water, metal, wood, and earth. A pot with white, red, and dark blue flowers incorporates, earth, fire, metal, water and wood.

Embrace: Landscaping and gardens create a softening and protecting embrace around the house.

Edges: Use meandering curves in all paths and bed shapes.

Energy: Encourage Chi to flow freely throughout your garden by eliminating dead limbs and plants, and by using curving lines in all installations to encourage Chi to meander. Chi's natural tendency is to do this, so it is desirable not to provide any obstacles. Installing a side gate or arbor invites Chi into the back and sides of the house.

Elemental Shapes: Fill out any shapes that are not either square, rectangular, triangular, or circular. This can be accomplished with plantings, garden beds, or paths.

Entrances: Create an en-trancing doorway to your garden and to your house. Chi enters through entrances.

Emotions: These are incorporated when you use symbols, spiritual

"We nourish the soul when we find value in the stillness of the moments, recognizing that the present time is the only time there is."

Gerald Jampolsky

111

practices, soulful considerations, plants that remind you of loved ones and pleasant experiences, and sensual treats.

HOW DOES FENG SHUI HEAL?

Feng Shui provides many ways to bring healing and positive energies to you and your family. I have worked with clients who had serious issues in their lives lifted after working with the Feng Shui enhancements that I recommended for them. Some consultations require reading the energy on the property, in the person(s), or in the home before a solution can be obtained. In any case, anxiety, stress, and worry are lifted, providing a way to find clarity and ease discomfort.

People report all the time how much better they feel in their homes or gardens after I work with balancing the energy. Very often it is as easy as moving a piece of furniture, removing a negatively charged item, bringing in colors that work to balance the five elements, or creating a curved bed where an angular one existed before. These corrections in energy flow make people feel more comfortable and nurtured by their surroundings rather than irritated by them. This causes relaxation and a renewed love for their space. Whereas before they described having an uneasy and uncomfortable feeling, now they report how wonderful it is to come home or go to their gardens. Their lives seem to work better and relationships improve.

Healing occurs because blocked, stagnant, or negative energy has been removed. Feng Shui provides ways to look at a problem's many layers and uses solutions that appear magical, yet are really only working with the energies of the Earth, people, and places. There is abundant proof that Feng Shui makes people feel supported, happier, more prosperous, and more in control of their lives. This reduces stress and anxiety, both producers of disease and discomfort.

Feng Shui heals the planet when it creates balance and harmony in the lives of people which then filters down to their family, friends, community, and the rest of the world. Feng Shui heals the Earth when it creates balance and harmony in gardens and landscapes and teaches regard for the sacredness of all living things.

The information I have provided in these Feng Shui chapters gives you a basis

"Soul is elicited from the things of the world by using them in a sacred way."

Eliezer Shore

from which to begin using these concepts in your garden. Please keep in mind that this is only a start; there is much to learn in Feng Shui. It is, after all, a 3,000-year-old tradition. Coming into contact with the West, it has changed and expanded somewhat. That's good, in my opinion. I have presented the views of the Black Hat School of Feng Shui with some additional eclectic ideas. Reading through these chapters, I am sure you have noticed that a great deal of common sense is used in every enhancement. The Compass, Form, Pyramid, and Western Schools of Feng Shui agree with many of these concepts and solutions. I generally follow an eclectic approach in Feng Shui because each school has much to offer.

The essential spiritual nature of Feng Shui brings soul into every aspect of theory, application, and thought. Beauty and aesthetics are of prime importance in Feng Shui, as is concern for the environment and applying sacredness to every act. Feng Shui is an excellent vehicle for the creation of soulful environments in your garden, landscape, or home.

> *"Compassion equals soul kindness. So, showing compassion for nature helps us treat our soul with kindness."*
>
> John Gray
> author of *Women are from Venus, Men are from Mars*

🍂 🍂 🍂 *Garden Gift:* 🍂 🍂 🍂
SEED PACKET GIFTS

At the end of the season gather seeds from your prized herbs and flowers, put them in bags, tie them with a ribbon and give as gifts from your heart.

PART THREE

A NEW VIEW
OF THE EARTH

"The Old Lakota was wise. He knew that
man's heart away from nature becomes
hard, he knew that lack of respect for
growing, living things soon led to lack
of respect for humans too."

—Lakota Chief, Luther Standing Bear

CULTIVATING HARMONY BETWEEN SOUL AND SOIL

"The way to an ecological life is to treat our home as the entire planet."
—Thomas Moore

Our environment is a fragile ecosystem of interconnected life forms. We have become accustomed to chemicals and their effects without giving thought to our future and our children's future. We pretend little is wrong with our world because we need the products of industry to sustain our so-called 'standard of living.' I have included in this chapter some of the many disastrous consequences of our oversight; there are thousands more. I have never been the same since reading *Silent Spring*, by Rachel Carson more than 41 years ago; it changed my views of Nature and our society. I wholeheartedly recommend reading this book as a place to start understanding how pollution works to affect everything and everyone.

THE HISTORY OF SOUL IN OUR CULTURE

In the Middle Ages, Jewish, Muslim, and Christian theologians adopted Aristotle's view that the human was a higher form than all other Earth species. Many centuries later, they translated their goals

"Gardening is an active participation in the deepest mysteries of the Universe."

Thomas Berry

117

into a need to control Nature, influencing medicine, the sciences, education, philosophy, the arts, and social and cultural life worldwide.

As desire for power and control grew, many social institutions began to ignore the soulful considerations of life and living. This encouraged the very attitudes and practices that helped to create our present disregard for the Earth. In time, western society lost track of the spiritual and soulful qualities of all life forms and became dedicated to a materialistic and mechanistic worldview. Science and technology have secularized our culture, affecting every aspect of daily life by removing soul from society.

René Descartes, a 17th century French philosopher, gave credence to the theory that body, mind, and soul were separate entities and viewed man as a machine that could be divided into parts, studied, and defined. Previously, body, mind, and soul represented the whole person and were not considered separate entities as they were after Descartes. He stated that anything that cannot be proven must be doubted. He also followed Sir Francis Bacon's premise that the aim of science was "to render ourselves the masters and possessors of nature." Descartes' well-known statement, "I think, therefore, I am" created a new view of man which separated the mind from the body and, coincidentally, also the soul. This and other changes in critical thinking paved the way for the development of all modern science and technology.

I suppose it had to be done in order for modern industry to develop. In the process we lost our belief that elements of soul resided in Nature, science, art, all living forms, and life itself. The spiritual and soul concept of Nature has had several revivals over the years only to be forgotten and consumed by the overwhelming power and control of government, various social institutions, and industry. Many are now re-evaluating these theories and are beginning to examine what we are destroying as we bow to the seductive forces of science and technology.

The New Physics is asking questions about the divine aspects of Nature, thereby viewing the universe in an entirely new way. They are fusing metaphysics and some concepts of physics into one body of knowledge, experience, and science as it was before Descartes. Discoveries have proven there is intention and consciousness in all of Nature and this helps us describe the order and spiritual quality of the universe.

No longer are we intimately connected with the process of growing food and the hardships that entails. Food appears in our markets as if by magic and we give no thought to how it got there.

118

Spend your money in ways that support your values.

In, *The Dream of the Earth*, by Thomas Berry, we are asked to listen to what the earth is telling us. He urges us to move from an anthropocentric view to a biocentric view of the world. Why? Because industry has been imbued with mythic power and we are allowing it to destroy the land without vision, thought, or consciousness. "We are destroying modes of divine presence . . . and are turning the most luxuriant forests into throwaway products," says Berry. We have become alienated, arrogant, and disrespectful. Restoring the air, water, and land requires that we restore the mystery and magnificence of the wind, rain, and soil.

Our culture values the very things that contribute to our alienation from the Earth. We allow the destruction of rainforests to satisfy our desire for disposable paper products and encourage the use of automobiles that foul our air. We seemingly ignore the loss of fertility in our soils in spite of the overuse of fertilizer and insecticides and their pollution of our rivers. As this disregard mounts, the healing qualities of the Earth are forgotten because we have placed more importance on the material things of life than on the spiritual or eternal.

Industry has developed an adversarial attitude toward the environment. It is born of fear and a desire for power and control, so that things will go their way and profits will continue to rise. Nature is very often erratic. Industry finds it difficult to operate on chance. We can see firsthand the results of corporate greed as the scandals enveloping Enron, Anderson, Worldcom, tobacco companies, Monsanto, and Dow, to name a few, unfold in our living rooms. We don't hear much about those concerning the drug, oil, lumber, and other chemical companies; they exist as well. You and I are paying for these monetarily and spiritually. They will continue to occur, until we come together as one voice and demand another way.

"In a garden the process of soul-building and soil-building become one."

Christopher and Tricia McDowell

119

LAWNS AND GOLF COURSES

Homeowners and golf courses use millions of pounds of chemicals on their lawns yearly hoping to keep the bugs and weeds under control. This can sometimes reach the level of a military campaign, using up countless hours and dollars in the effort. Their lawns are veritable toxic dumps, devoid of all life. The soil under these massive lawns is poisoned, and runoff from these areas goes right into the local streams, eventually polluting the whole watershed. We have to drink this water downstream.

Nitron Industries is a company that has developed programs that help homeowners and golf courses use completely organic methods to care for their lawns, saving them time and money as well. (See Resources.) Soil is a miniature ecosystem that when cared for contains all the things healthy plants and lawns need.

Almost all of the chemicals used for lawn care have been linked to many forms of cancer, birth defects and decreased fertility, and they linger in the environment for years. These chemicals are highly estrogenic and have been implicated in the rise in breast and prostate cancers recently and also have been known to cause health problems in the workers that handle them.

I suggest trying to plan less lawn space around a home. Use areas of plantings, beds, walkways, paths, ponds, berms, rock gardens, and trees with plants around them to reduce the lawn you have to take care of to modest amounts. I am sure you would love to devote your time and money to more enjoyable garden pursuits. Gardening catalogs now recommend using ground covers. Creeping thyme, chamomile, blue star creeper, pachysandra, myrtle and others may substitute for a lawn. Many homeowners have opted for turning their labor-intensive lawns into wildflower prairie habitats. One can also intersperse patches of lawn with areas of assorted groundcovers. This provides a treat for the eyes, since they all have different textures, flowers, and colors. The choices are limitless.

One way to increase the levels of harmony between soul and soil is to buy organic products and support your local health food stores every chance you get. In the next paragraphs, I outline many of the reasons why I feel this is important and why it restores balance and soul to the world.

Remember when you eat food you are also ingesting air, sun, sky, rain, clouds, earth, and water.

120

VEGETABLES

Vegetables raised today do not have the same levels of nutrients they once did. The nutrient value of carrots, kale, broccoli, spinach, and others have been reduced by 50 percent over the last 20 years. Roman wheat sustained the Roman army because it had much more protein content than our present-day wheat.

Changes in broccoli's nutrients since 1975:	
Calcium:	down 53.4 percent
Iron:	down 20 percent
Vitamin A:	down 38.3 percent
Vitamin C:	down 17.5 percent.

Alex Jack, a macrobiotic nutritionist, had broccoli and other vegetables tested and was astonished to find that calcium had fallen by 27 percent, iron by 37 percent, vitamin A by 21 percent, and vitamin C by 30 percent. The Food and Drug Administration was unable to offer an explanation. Hasn't anyone told them it's a no-brainer? It's the soil.

ORGANIC FOODS

Organic foods are now a multi-billion-dollar industry. Dole is already selling organic bananas. Many people complain that organic foods cost more. I have found they last longer in the refrigerator because they are fresher, denser, and richer in naturally occurring preservatives. The aromatic and essential oils that are the soul of the plants help them stay fresher longer. They taste better, are more satisfying to eat, and even look better. Organic farmers take immense pride in what they do. They raise their produce with good intentions

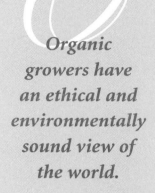

Organic growers have an ethical and environmentally sound view of the world.

121

and a spiritually conscious regard for the Earth. I am sure it also affects the food.

I also suggest doing the taste test by comparing the corn in your supermarket with those ears at the organic markets. The organic corn is considerably sweeter and more succulent. I pay 19 to 49 cents each for lemons at my local co-op. At the local super market they are the same price and don't taste as good, are as hard as rocks and aren't as juicy. I know I get more for my money and my body thanks me when I buy organically grown produce.

What if you paid a little more for organic foods? Would your budget absorb around $10 to $20 more a week? If your SUV was getting 40 miles to the gallon instead of 11 to 17 miles per gallon, you could convert those savings to pay for your food.

Research has revealed that 54 percent of Americans would like to see organic agriculture become the predominant form of food production. It is estimated that by the year 2010, 10 percent of American produce will be raised organically. In Europe trends indicate that 30 to 50 percent will be organic by 2010. People are now concerned about pollution, pesticide residues, food poisoning, irradiation, and biodiversity.

Community Supported Agriculture (CSA)

Another way to get high-quality produce for your family while supporting organic and local growers is to join a local community supported farm.

These farms ask that you sign an agreement with the farm or cooperative to buy a certain amount of produce each week for the growing season. Usually there is a newsletter with farm news and recipes. Subscribers pay from $60 a month to $500 a year and pick up their bounty each week at a designated spot. Most farms are family run. Some supply crops to local restaurants and stores, and to the less fortunate through meal programs. By buying from them you are ensuring the future of family owned farms in the United States. There are approximately 1,000 in the United States and Canada at this time and the number is growing.

Sustainable agriculture yields high-quality plants and does not deplete or damage soil, water, or wildlife, promotes the health of the environment, and supports biodiversity.

$\mathscr{B}$ECOMING AN EARTH STEWARD

"We are beginning to understand that the quality of our relationship to the Earth will determine the quality of our future."

—Jeff Cox

If we are to flourish and leave a healthy planet to future generations, we must reassess the way we grow, package, and distribute food. This chapter provides some examples of projects changing the way cities, towns, neighborhoods, and businesses view the environment and gardening, as well as educate future generations. As destruction of the environment increases, we find pockets of enlightenment; people are taking responsibility for the environment, changing their attitudes and practices and spreading awareness that gardening organically and with soul can save the Earth. Landscaping, farming, and gardening organically "not only heals us, it heals Nature too," says Jeff Cox.

EARTH STEWARDING PROJECTS

Chicago "Urban Re-Leaf Program"

Chicago is hoping to convert urban sprawl into urban green, to clean the city air and lower hot summer temperatures by planting trees and gardens on the roofs of downtown buildings. Studies have

"As we assault Nature around us we assault Nature within us and vice versa."

Michael Cohen

GARDENS OF HOPE: The Gateway Gardening Organization was formed In St. Louis, Missouri because research showed that oases of green can improve communities. This organization helps people find plots in the city to start a garden and prevents present plots from being turned over to development. In addition to less crime, and less graffiti, people are outside more, which encourages them to develop more interest in their community and in one another. It nourishes people right where they live.

shown that in some cities the tree cover is down 15 percent due to disease and the paving over of green spaces for development. Trees absorb and deaden city noise and take up sewer overflows. As the water is transpired and evaporated from their leaves the surrounding air is cooled and cleaned of carbon dioxide and the oxygen content of the air is improved. *Country Living*, September/October 2001.

Farming

A third generation farmer from Kansas interviewed by Bill Moyers on the PBS program "Now" reported he has turned his farm over completely to organic methods because he has found they are cheaper and are not as energy intensive as non-organic methods. His plow makes only strips for the seeds, saving time and energy, and keeping soil from blowing away. He does not have to spray for bugs or use herbicides that raise the cost of producing crops. According to the latest information, he can reduce his energy expenditure 50 percent by implementing these changes. Before 1940 farmers lost 33 percent of their crop to insects without the use of chemicals. The present loss is also at 33 percent with the use of chemicals.

Schools and the Environment

Many schools are becoming interested in composting their cafeteria wastes to use in their landscaping and in teaching their students the valuable lessons of recycling, conscious living, and organic processes. A restaurant in Berkeley, California, has started what they call the "Edible Schoolyard," behind the gym of a school in their community. Focus will be on a compost pile, fed by scraps and waste from the school cafeteria, an adobe bread oven, an orchard of espaliered fruit trees, vegetable beds, herb borders, and a creek with rocks. This approach will provide a fertile learning environment right behind the school. (edible@lanminds.com)

Green Solutions for Gardeners

1. Turn on watering systems early in the morning or early evening. This saves precious water from evaporating and lowers your bills.

2. Use native plants. They require less water.

3. Put in ground covers instead of high-maintenance lawns.

4. Try to use fewer insecticides and herbicides.

5. Plant more trees to cool your home and use the AC less.

6. Mow leaves in the fall and leave them on the lawn for mulch and nutrients.

7. Use solar-powered lights outside when you can.

8. Use a push mower to save gas and reduce air pollution.

9. Cut your grass not lower than 3 inches; taller grass shades the soil, and reduces surface heat and evaporation so your lawn will need less watering.

10. Start a compost pile so you don't use the landfill as much.

Household EcoTeam Workbook and Program

This program has developed techniques for homeowners to reduce garbage and home water use and improve energy and transportation efficiency while at the same time encouraging eco-wise consumer habits and empowering others to do the same.

An article in *Family Circle*, March 13, 2001, discusses how a family that followed the plan described above learned that conserving paid off for them not only in more responsible and thoughtful

"Mono-cropping depletes the soil and sets it up for pests, etc."

Roger Yepsen

choices but savings in dollars as well. The small changes they made had a huge impact on their life as a family. The parents feel it taught their children more respect and appreciation of the environment as well as creating a greater bond of cooperation and purpose between family members. The EcoTeam program and workbook gives you a place to start to implement changes that will affect you and the environment favorably.

If a family of four takes these actions, in one year 10 trees, 73,000 gallons of water, 104 cubic feet of landfill, 3,120 pounds of garbage, 600 gallons of gasoline, and $1,200 will be saved. In addition 140 pounds of acid rain will not fall on our forests and 10 tons of carbon dioxide emissions will be eliminated.

(Each car uses 3,000 gallons of gas in its lifetime and spews out 35 tons of carbon into the air adding to the already endangered ozone layer and increasing planetary temperatures.)

Things You Can Do in Your Home to Help the Environment

- Avoid toxic oven cleaners.

- Try Bon-Ami, borax, white vinegar, or baking soda for most household cleaning chores instead of harsh chemicals.

- Hydrogen peroxide or white vinegar may be used as disinfectants. The combination of the two kills 99 percent of bacteria. (From *Alternative Medicine*, January 2002, page 22.)

- Use saved rags instead of paper towels.

- Explore the world of new paper products that do not use chlorine bleaches.

- Use a clothesline when you can.

- Don't let the water run while doing the dishes or brushing your teeth.

A book to look into: *The Naturally Clean Home*, by Karen Siegel-Maier.

"Is our society ecocidal?"

Pamela Jones

THIS QUOTE OFFERS US WAYS TO BECOME AN EARTH STEWARD.
THE VALUE OF NATURE

by U.S. Representative Jay Inslee, Washington State
(*Sierra*, The Magazine of the Sierra Club, January/February 2002, with permission from Representative Inslee)

In nature we find peace, even in the face of war. A house finch taught me this lesson. At the end of a long week in Congress last September, I sat in my Washington, D.C. apartment watching the horrific images of man's madness on television for the thousandth time. I was despairing of someday gaining a lasting peace, when I happened to glance up and see a house finch serenely perched in a maple tree at my window.

That little bundle of feathers took me away from the darkness of the week's horror. He was thinking about seeds and lady finches, not terrorist cells and exploding planes. It was a comfort to be so close to a creature that could sit enjoying the morning sun, and greet it with a song—things I was incapable of at the moment. If I could have thanked him for the minute's respite, I would have.

We need nature for ourselves, not just for itself—perhaps more than ever during the trials of war. In this time of new personal and public priorities, we must continue to work to protect nature. Places like the Arctic National Wildlife Refuge should not fall victim to the terrorist attack on America. It is true that our nation should shift its energy policy, but not from one hole in the ground to another in our search for oil. Just as a man addicted to cigarettes can save his life by quitting smoking—but not by switching brands—we can save ourselves by increasing conservation and use of alternative fuels. We should reduce our addiction to oil from any source.

It wouldn't be difficult. We could set vehicle mileage standards for SUV's and light trucks at 40 miles per gallon and save 8 times as much oil as is economically recoverable from the Arctic National Wildlife Refuge. We could invest in wind turbines, which are so efficient that Denmark has decided to produce 50% of its electricity in this manner by 2010. We could work toward a hydrogen-based transportation system, as Iceland is doing. We could re-examine solar power technologies, which have made great strides over the past decade.

There is a connection between nature, energy, and war—between the golden plovers that migrate from the Arctic Refuge and the societal troubles in foreign countries that migrate to America's shores. Nature can help provide us with a kind of peace that lasts for more than a moment. If our intense feelings for the wildlife of the Arctic inspire us to wean ourselves from oil, we will certainly enjoy more nature and more peace. Henry David Thoreau said it grandly, "In wildness is the preservation of the world." A house finch said it simply.

127

These Are Things I Have Been Doing to Steward the Earth.

I write letters, send e-mails, and make calls for environmental groups. I have joined and contribute to ecology-based organizations and remain vigilant and informed about our environment while supporting local recycling efforts. Whenever weather permits, I dry my clothes outdoors to save energy and to experience that wonderful aroma which permeates my laundry. My garden and yard are herbicide and insecticide free and I have eliminated chlorine-based products from my home. I support my local co-op, health food stores, and farmers markets. We all need a long-distance phone carrier. I use Working Assets (866-225-9253), a long distance telephone company that supports the environment and gives millions each year to worthy social and environmental causes. They do this by asking their customers to round out their bill every month.

It's hard to tell the truth when it seems the whole world is against what you see. On most occasions I advocate reporting when "The Emperor has no clothes on." I agree, it would be difficult, and in some cases downright dangerous, to air controversial views about what industry is doing to the environment. There is safety and power in numbers, so I advise joining environmental groups and working with them; they tell the truth. The ecology and conservation movement believes that everything alive on this planet deserves respect and care.

Ecology at its deepest level is the spiritual practice of making and finding a home for the soul while at the same time caring for the Earth. Separating the word ecology into its parts of *eco* and *logy* we find that in Greek *eikos* means our emotional search for home and *logos* is the study of or implying the full mystery of God. The experience of home relates not only to shelter but feeling protected and enlivened by the natural world and knowing that the entire planet is our home.

There are many ways that we can increase the stewardship of our precious resources. In doing so, we nourish our soul in the present and make sure the Earth will prosper and thrive in the future. It is important to be vigilant in the same way we preserve our democratic way of life. I especially remember Representative Inslee's petition for the future. His words inspire and motivate.

I am convinced that the depletion of resources occurring all over the planet can be halted by the concern and increased sense of responsibility of gardeners whose daily work protects the bond we have with Nature. It is important for all of us to step forward and be accountable. We have an obligation to future generations. Human beings are powerfully drawn to the Earth and to gardens. There is an established connection between humans and plants. As industry threatens to change our climate, destroy our source of oxygen, deplete our natural resources and supplies of clean water, and pollute the air and soil, we have an opportunity to create change from the utter chaos.

As more and more people begin to care for the Earth, companies will notice what consumers are buying. Soon, they will offer more products to satisfy those desires. It is an interconnected chain that flows from you to garden supply companies back to you and then to the environment.

You can start in a small way by turning your lawn over to groundcovers, or you can decide to use only organics on your vegetable beds. Maybe you will start to use only organic methods to work with insects and weeds. Whatever you choose to do will change you in subtle ways. Your change will affect other people. Your neighbors will notice the changes. Your family will start to make other soulful choices. Finally, you will have a positive effect on your community and the world. It's that easy.

Your simple act, your simple choice, can affect millions and most importantly, the Earth. By buying organic products, you will have supported what Lester R. Brown, founder and president of Earth Policy Institute, calls the "eco-economy." He writes in *The Natural Business LOHAS Journal,* summer 2002, "By far the most urgent need is to get the market to tell the truth—to tell the ecological truth. Now, we have a situation where if you buy a gallon of gasoline, for example, you do not pay the costs of increased healthcare expenditure for treating respiratory illnesses from breathing polluted air caused by burning gasoline. We do not pay the costs of acid rain damage. We do not pay the costs of disrupting the earth's climate system. We need to get all these costs on the table, to get the market to tell the ecological truth."

129

LOHAS Journal is dedicated to the promotion of sustainable business as a way to fundamentally alter the landscape for economic, social, and environmental change.

A Civilization that Makes Sense

"What do we need to know to make sense out of our lives? To make sense as individuals? As a society? As an endangered species? Exactly what is the crisis we face? What can we do to solve It?

When we can answer these questions we will achieve a new kind of literacy . . . a literacy that goes beyond the definition we presently use . . . beyond mastery of the basics we now consider a 'good education.' A growing number of scientists and scholars tell us that mastery of a new level of literacy is essential. Without it, we can't create a civilization that makes sense . . . human sense . . . planetary sense . . . cosmic sense. The new literacy we must master is Earth Literacy."

—from materials printed by The Environmental Ethics Institute, Miami-Dade Community College.

We can restore soul to the world. We can begin by finding it in our gardens. If each one of us does one or two things outlined in this chapter, it could change some dangerous situations currently impacting our environment. Viewing the inspiring scenes that Nature provides so readily and with such gusto, we can only come to one conclusion; re-thinking our economic and social priorities is a necessity.

You have the power of choice. Your shopping choices speak volumes to the government, the media, and industry. You can help restore balance and soul to our fragile world by choosing to be an Earth steward. Go for it.

RESOURCES

BIBLIOGRAPHY

WATER

Alexanderson, Olaf. *Living Water: Victor Schauberger and the Secrets of Natural Energy.* Dublin, Ireland: Gateway Macmillan, 1982.

Coats, Callum. *Living Energies: The Schauberger's Work with Trees, Light, Air, and Water.* Dublin, Ireland: Gateway Macmillan, 1996.

Emoto, Masaru. *The Message from Water.* Kyoikuysha, Tokyo, Japan: HADO, 2000.

Froelich, Jacqueline and Barbara Harmony. *AquaTerra: MetaEcology and Culture.* Eureka Springs, AR: The Water Center, 1995.

Ryrie, Charlie. *The Healing Energies of Water.* Boston, MA: Journey Editions, 1999.

Schauberger, Victor. *The Water Wizard: The Extraordinary Properties of Natural Water.* Dublin, Ireland: Gateway Macmillan, 1998.

Schwenk, Theodor. *Water: The Element of Life.* Hudson, NY: Anthroposophical Press, 1989.

Water: The Drop of Life. PBS Series. October, 2000.

SOUL

Andrews, Ted. *How to Work with Spirit Guides.* St. Paul, MN: Llewellyn Publications, 1997.

Carslon, Richard and Benjamin Shield, Ph.D., *Handbook for the Soul.* Large Print Edition, New York, NY: Little, 1999.

Clinebell, Ph.D., Howard. *Ecotherapy.* Minneapolis, MN: Fortress Press, 1996.

Cohan, Michael. *Reconnecting with Nature.* Corvalis, OR: Ecopress, 1997.

Couisineau, Phil. *Soul Archeology.* New York, NY: Harper, 1995.

Goldman, Connie and Richard Mahler. *Tending The Earth, Mending The Spirit.* Center City, MN: Hazelden, Publishing, 2000.

Lawlor, Anthony. *A Home for the Soul*. New York, NY: Charles Potter, 1997.

Marcus, Claire Cooper. *House As a Mirror of Self*. Berkeley, CA: Conari Press, 1995.

Moore, Thomas. *Care of the Soul*. New York, NY: Harper Perennial, 1992.

Moore, Thomas. *Education of the Heart*. New York, NY: Harper, 1997.

Moore, Thomas. *The Re-Enchantment of Everyday Life*. New York, NY: Harper 1996.

Search for the Soul. Alexandria, VA: Time/Life Book, 1989.

Stone, Jana, Editor. *Every Part of The Earth is Sacred. Native American Voices in Praise of Nature*. New York, NY: Harper, 1993.

Zukav, Gary. *Seat of the Soul*. New York, NY: Fireside, 1990.

THE SENSES

Abram, David. *The Spell of the Sensuous*. New York, NY: Random House, 1997.

Ackerman, Diane. *The Natural History of the Senses*. New York, NY: Vintage Books, 1991.

Cox, Jeff. *Creating a Garden for the Senses*. New York, NY: Abbeville Press, 1993.

Eiseman, Leatrice. *Color for Your Every Mood*. Washington, D.C.: Capital Books, 1998.

Malitz, Jerome and Seth. *Reflecting Nature*. Portland, OR: Timber Press, 1998. (Inspiring nature photography)

ECOLOGY

Berry, Thomas. *The Dream of the Earth*. San Francisco, CA: Sierra Club Books, 1988.

Carson, Rachel. *Silent Spring*. New York, NY: Houghton Mifflin, 1962.

Earth Works Group. *50 Simple Things You Can Do to Save the Earth*. Berkeley, CA: Earthworks Press, 1989.

Fukuoka, Masnobu. *The One Straw Revolution, An Introduction to Natural Farming*. Emmaus, PA: Rodale Press 1978.

Jones, Pamela. *Just Weeds, History, Myths & Uses*. New York, NY: Prentice Hall/Simon Schuster, 1991.

Olkowski, Helga and William. *The Gardener's Guide to Common Sense Pest Control*. Newtown, CT: Taunton, 1996.

Pfeiffer, Ehrenfried. *Weeds and What They Tell Us.* Kimberton, PA: Bio-dynamic Farming and Gardening Association, 1950.

Rodale, J.I. and Staff, *The Complete Book of Composting.* Emmaus, PA: Rodale Press, 1967.

Roth, Sally. *Weeds, Friend or Foe? An Illustrated Guide to Identifying, Taming, and Using Weeds.* Pleasantville, NY: Reader's Digest Association, Inc., 2000.

Tomkins, Peter and Christopher Bird. *Secrets of the Soil.* New York, NY: Harper & Row, 1989.

Wallace, Dan, Editor. *The Natural Formula Book.* Emmaus, PA: Rodale, 1982.

Yepsan, Roger B., Editor. *Organic Plant Protection.* Emmaus, PA: Rodale, 1976.

FENG SHUI

Hale, Gill. *The Feng Shui Garden.* Pownal, VT: Storey Communications, 1998.

Lin, Jami, Editor. *Contemporary Earth Design.* Miami, FL: Earth Design, 1998. Terah Kathryn Collins, "Outdoor Feng Shui" p.155-164.

Linn, Denise. *Feng Shui For the Soul.* Carlsbad, CA: Hay House, 1999.

Kennedy, David Daniel. *Feng Shui For Dummies.* Indianapolis, IN: John Wiley & Sons, 2000.

Too, Lillian. *The Illustrated Encyclopedia of Feng Shui.* New York, NY: Barnes and Noble, 2000.

Webster, Richard. *Feng Shui in the Garden.* Minneapolis, MN: Lewellyn, 1999.

Wydra, Nanci Lee. *Feng Shui in the Garden.* Lincolnwood, IL: Contemporary Books, 1997.

GARDENING WITH SOUL

Ackerman, Diane. *Gardening Delight.* New York, NY: Harper, 2001.

Handelsman, Judith. *Growing Myself-A Spiritual Journey Through Gardening.* New York, NY: Dutton, 1996.

Keller, Debbie. *The Spiritual Garden.* Kansas City, MO: Ariel Books/Andrews McMeel Publishing, 2000.

Lacy, Allen. *The Inviting Garden.* New York, NY: Henry Holt, 1998.

McDowell, Christopher Forrest and Tricia Clark-McDowell. *The Sanctuary Garden.*
New York, NY: Simon and Schuster, 1998.

Rawlings, Romy. *Healing Gardens.* Minocqua, MN: Willow Creek Press, 1999.

The Findhorn Community. *The Findhorn Garden.* New York, NY: Harper & Row, 1976.

Wright, Machael Small. *Garden Workbooks I & II*, Warrenton, VA: Perelandra, 1986.

OTHER TOPICS

Broomfield, John. *Other Ways of Knowing.* Boston, MA: Tuttle, 1997.

Chard, Philip Sutton. *The Healing Earth.* Minnetonka, MN: North Word Press, 2000.

Elgin, Duane. *Voluntary Simplicity.* New York, NY: William Monroe & Co., 1993.

Editors of Time-Life Books. *Earth Energies.* Alexandria, VA: Time-Life Books, 1991.

Gerson, David and Gilman, Robert. *Household EcoTeam Workbook.* Olivebridge, NY:
Global Action Plan, 1991.

Goelitz, Jeffrey. *Secrets from the Lives of Trees.* Boulder Creek, CA: Planetary Publications, 1991.

Gregg, Richard and Philbrick, Helen. *Companion Plants and How to Use Them.* Old
Greenwich, CT: The Devin-Adair Co., 1966.

Kite-Friedman, Pat, *Bird Lovers Garden.* Hales Corners, WI: Fairfax Publishing, 2000.

Lonegren, Sig. *Labyrinths, Ancient Myths and Modern Uses*, Glastonberry,Somerset,
England: Gothic Image Publications, 1991.

Shaudys, Phyllis, *Gifts from the Garden.* Pownal, VT: Storey Communications, *1990.*

Simpson, Liz. *The Healing Energies of the Earth.* UK: Gaia Books, 2000.

Tompkins, Peter and Christopher Bird. *The Secret Life of Plants*, New York, NY: Harper
Trade, 1989.

Whitner, Jan Kowalczewski. *Rock Talk-StoneScaping: A Guide to Using Stone in Your
Garden.* Pownal, VT: Storey Communications, Inc., 1992.

MAGAZINES

The Natural Business LOHAS Journal
Fine Gardening

Organic Gardening

Kitchen Garden

Country Living

Family Circle

Water Gardening

Pangala, R.J. Stewart, "Faeries"

The Herb Quarterly, spring 1999, "The Medicinal Berry Patch," Barbara MacPherson.

Parabola, summer, 1996. "Through A Dark Passage," Eliezer Shore.

Acres USA, 800-355-5313, Eco-Farming/Eco-Living.

ASSORTED GARDEN WEBSITES

www.Labyrinthproject.com

www.sacredspaces.com

www.watergarden.com

www.ponderingthoughts.com

www.watergardening.com

www.duncraft.com (birds)

SOURCES FOR ENVIRONMENTALLY FRIENDLY GOODS & INFORMATION

www.earthshare.org, resource of environmental information

www.biodynamics.com, supports biodynamic agriculture practices

www.realgoods.com, environmentally safe products

www.gaiam.com, environmentally safe products

www.seventhgeneration.com, environmentally safe products

www.lohasjournal.com, promotes environmentally sound business practices

www.washingtonfreepress.org, a good source of environmental information

www.organicvalley.com, a consortium of organic growers

www.smartwood.com, environmentally sound wood products

www.buildinggreen.com, environmentally sound building products

www.ecotimber.com, ecologically sound timber practices
www.greenpressinitiative.org, environmentally sound uses of paper
www.birc.org, The Bio-Integral Resource Center
www.attra.org, solutions for every gardening problem, 800-346-9140
www.members.aol.com/rccouncil/ourpage, Rachel Carson Council Inc.
www.beyondpesticides.org, National Coalition Against the Misuse of Pesticides
www.nationalwatercenter.org and www.planetaryhealer.net, The National Water Center.

ENVIRONMENTAL ORGANIZATIONS
www.citizen.org, Public Citizen
www.organicconsumers.org, Organic Consumers Association
www.sierra.org, Sierra Club
www.NRDC.org, National Resources Defense Council
www.fscus.org, Forest Stewardship Council
www.bioneers.org, Bioneers

ENVIRONMENTALLY FRIENDLY WEBSITES AND SOURCES
www.groworganic.com, 888-784-1722 (garlic barrier insect repellent.)
www.gardeners.com and 800-427-3363, Gardeners Supply Company offers
    many products for insect problems.
www.arbico.com, complete source of environmentally sound products
    for insect control.
www.nitron.com, Nitron Industries, 800-835-0123. Offers everything
    organic for the garden.
www.chelseagreen.com, Chelsea Green Publishing Co.
    (Published "The Man Who Planted Trees".)
www.rodale.com, publisher of many environmental books.
Wildwood Farms, Inc., P.O. Box 939, Clinton, IA. 52733.
    (information about squirrels.)

Findhorn Foundation, Forres, Scotland IV 36 ORD.
Email: Engarden@aol.com, Leslie Goldman

SEEDS AND PLANTS

Seed Savers Exchange, 3076 N. Winn Road, Decorah, Iowa, 52101.
The Grain Exchange, 2440 Water Well Road, Salina, Kansas, 67401
Native Seeds/Search, 2509 N. Campbell Avenue,
   #325, Tucson, Arizona, 85719. 520-327-9123.
www.johnnyseeds.com
www.selectseeds.com (antique flower seeds)
www.orderseed.com, Territorial Seed Company
www.wayside.com
www.dutchgardens.com, 800-818-3861

DANDELION REFERENCES

Hoffman, David. *The New Holistic Herbal.* Brookline, MA: Redwing Books, 1992.
Wardwell, Joyce. *The Home Herbal Remedy Book.* Pownal, VT: Storey
   Communications, 1998.
Herbal Voices Newsletter, Joyce Wardwell, publisher, 310 Mt. Bliss Road, East Jordan,
   Michigan, 49727, 616-536-2877.

ENVIRONMENTAL ISSUES

INSECTICIDES

Each year more than 4 billion barrels of insecticides are used worldwide and 90 percent of them have never been tested for long-term health effects. Over 2 million people suffer from insecticide poisoning worldwide yearly, resulting in 40,000 deaths as of 1990. Children receive a greater exposure than adults because of their consumption of food, air, and water per pound of body weight. One study showed that in households where insecticides are used children are much more likely to suffer from leukemia, which is a growing medical concern in children.

CHLORINE

The processing of paper products is accomplished with chlorine, which bleaches and softens the pulp. It turns into organochlorines that have been implicated in lowered IQ, reduced fertility, genital deformities, breast cancer, prostate cancer, reduced sperm counts, and abnormalities of the immune system. In many household cleansers it masquerades as sodium hypochlorite. (From Seventh Generation.)

You can protect yourself by reducing the amount of paper products that you use near you skin, in or on your body, or for household cleaning. Make an effort to purchase paper products from a health food store. I know they are much more expensive yet you will be safe using these products. They are manufactured with ingredients that are not harmful to you such as hydrogen peroxide and sodium hydrosulphite. (See Seventh Generation and Real Goods catalog listings in Resources.)

LUMBER

The Home Depot and Lowe's have changed their policy of buying woods from endangered forests and Third World countries. They now stock certified lumber that is not from these sources.

Conservation and ecology groups increase the public awareness of clear-cutting methods that use gigantic machines to tear up the trees and build roads in National Park lands. Our National Park Service still bows to the pressures of big business, poor government planning, and outright unwise practices. If we desire to ensure having forests to take our grandchildren to, we all need to take an interest in these issues. The choice we make at our local home building suppliers, lumber companies, and hardware stores will help change the present attitude of leniency toward lumber companies robbing us of our future.

SLUDGE

Industry and citizens pour paint and chemicals right down their drains that eventually end up in the water supply and in the sludge that is left over after sewage is treated by municipal waste treatment plants. It becomes dangerous when it is sold to farmers who then spray it on their fields and use it as a fertilizer. It is also sold to gardeners with labels citing it as pasteurized, sanitized, reformulated, or some other euphemism that masks the sludge origins. The Environmental Protection Agency has tried to help municipalities and companies market their sludge products by dreaming up other euphemistic labels, such as "Biosolids." There is ample proof that dangerous chemicals such as mercury, lead, arsenic, cadmium, aluminum, asbestos, and radioactive and nuclear by-products remain in the sludge, become part of the soil mix, and then transfer to us through the food chain. (See *Fatal Harvest: The Tragedy of Industrial Agriculture* by Andrew Kimbrell.) Birds flying overhead, insects, little animals, butterflies and worms are all affected by this chemical concoction. The chain of life is interconnected, interwoven, and complex. Some of these very chemicals are used in the manufacture of computers and are now ending up in landfills as well.

Many of these chemicals have been linked to cancer, birth defects, decrease in fertility, and more and they linger in the environment for many years. DDT and Dursban are now banned, although there are many other products sold that are damaging our environment and causing health problems to the people who use them. Most of these products were developed for commercial agriculture applications, and home gardeners have now accepted them.

GENETICALLY ENGINEERED FOODS

Genetically engineered foods, Frankenfoods as Europeans call them, are being promoted and used by all the major food manufacturing companies in the United States. The Federal Drug Administration never approved these food additives. They are looking the other way while consumers are being used as test cases for this new technology. Food manufacturers hope to so completely flood the food market that consumers won't be able to find a single food without a genetically engineered ingredient and then won't have any choice.

SOLAR

Harnessing the sun at the University of Arkansas, Fayetteville, Arkansas.

Jack DeVore and Jim Snow, faculty members of the University of Arkansas, have demonstrated that energy is neither scarce nor expensive, by inventing a new way to cure lumber that harnesses the energy of the sun, along with a revolutionary design for solar collectors and drying lumber.

Their design allows lumber to go in and come out cured, dried, and ready for use without the 20 percent loss of board feet that now occurs in kiln drying methods. This could mean phenomenal savings to industries, consumers, forests, and the environment. There is no energy cost for using their solar collectors, which are curved, and also no loss of wood. The usual drying process creates a loss of lumber due to warping of the boards. Their invention can also be used to heat homes and water and can be applied to industrial use. They have applied for a patent.

The Morning News, NW Arkansas, August 20, 2001.

INDEX

AUTHOR BIO:

Gaylah Balter grew up using acorns as dishes and talking to the massive oaks under which she played. Sitting with her dog Pal on the hill outside her home she chatted with him about all sorts of things. She used to pick the irises and lilies of the valley that grew on the property where she lived, and she talked to them as well. It seems she was always talking with some part of the natural world.

As an adult, Ms. Balter traveled to many faraway places and lived the ordinary life of most people, marrying, working, raising two children, divorcing, and discovering more about herself. Gaylah became an occupational therapist, educator, meditation teacher, and massage therapist, specializing in energetic techniques for healing.

She has now settled in Fayetteville, Arkansas, and has what she calls her 'English Cottage' with a bay window in her kitchen looking out onto the expanse of her garden and pond. She does this quite often with her treasured dog, Joseph. In reality it's a little house on a little lot right behind a shopping center, yet to listen to her you would think she lives far out in the country.

Throughout her adult life, Gaylah has been active in social action and environmental organizations and has participated in numerous projects to lobby governmental agencies and increase public awareness about important Earth issues.

Gaylah has written two books since moving to Fayetteville. Her first was *Clean Your Clutter, Clear Your Life: A Practical Manual Using Feng Shui Principles* and her second book is *Gardening with Soul; Healing the Earth and Ourselves with Feng Shui and Environmental Awareness*. She is a Feng Shui consultant and teaches classes on meditation, Feng Shui, cleaning clutter, six healing sounds, and gardening with Feng Shui.

ARTIST BIO:

Gloria Pendry grew up on a farm in Iowa and since childhood had a desire to record the shapes and colors that attracted her eye. She says, "There is so much beauty all around me and in every place that I have been on my travels. It seems that I always have more subject matter than time."

She started taking art classes in the 1970s, and did much more drawing, and learned more techniques, until she finished her B.A. at the University of Arkansas. She and her husband have chosen to retire in Northwest Arkansas.

Many of her paintings are done in watercolor, as well as oils or acrylics. Gloria has also tried her hand at woodcarving, jewelry, various needlecrafts, and even tries to write a poem now and then. She worked in graphics for a few years, where she learned the pen and ink method used in this book.